Contents

Preface

Writing this book over the past three years has been like the proverbial painting of a long bridge: by the time one end is reached, the other needs redoing. But unlike a bridge, it has got longer each time. Not only is more and more being understood and written about rural poverty and rural development, but the act of writing has also forced me, not always willingly, in unexpected directions. The book began with six chapters but the last one grew into three, and now there is yet more I would like to add. But the time has come to stop, recognising this as only a stage in a journey in which there is far still to go.

The book is for people who are concerned with rural poverty and rural development. Some are from rich countries, but the great majority are professionals in Third World countries, working in government departments, voluntary agencies, political parties, commercial organisations, schools, universities, training institutes and research organisations. It is an attempt to speak to both practitioners and academics, and to both social scientists and physical and biological scientists, without distinction of profession or discipline. It is about rural poverty and the perceptions, attitudes, learning, ways of thinking and behaviour of professionals. Its original title was 'Putting the Last First: Reversals for Rural Development', and it retains reversals as a central theme – the need for them, their feasibility, and their personal implications.

The focus is deliberately limited to rural poverty and to the Third World. There is appalling urban poverty in the Third World, and there is rural poverty in the richer worlds. Some of what I have written applies to these, but rural poverty in the Third World deserves special attention and efforts because it is less visible. Within the Third World, much of the evidence is from Africa South of the Sahara and from South Asia; and while this no doubt influences the analysis and conclusions, I hope that what is said will be found relevant and useful in Latin America, the rest of Asia, and elsewhere.

I have restricted the subject matter in two further ways. First, I have not presented case studies or detailed analyses of programmes and projects which seek to reach and help poor people. Careful evaluation and comparison of such initiatives is a continuing need but it is a huge topic and deserves separate treatment. Second, I have limited the discussion to deprivation which is material and social. This is something which outsiders and the rural poor can agree in saying no to. But the material and the social are not the whole of life. There is also the spiritual side and the quality of experience and being. For those who have a decent and secure livelihood, the relationship between more wealth and greater happiness is an open question. For those at the lower end of the scale, trapped in poverty, things are clearer. Extremes of material and social deprivation can narrow awareness and warp, embitter and kill. So it seems all the more right to concentrate attention on the 'last', on the hundreds of millions of largely unseen people in rural areas who are poor, weak, isolated, vulnerable and powerless. Whatever one's ideology it seems right to reverse the forces which exploit these people and make and keep them physically and socially wretched.

There are, too, the limitations of the author. Most of these will be transparent enough. However, much I may try to see things in some universal or detached way, I cannot do so; and I cannot help being inside the skin where I find myself, as it happens that of a puzzled and uneasy middle class Englishman who has lived and worked in Africa and Asia. Nor have I been able to resist, despite good advice, occasionally having fun with language. All I ask is that the reader should not discount the evidence, ideas and arguments of the book for these reasons, but rather accept or reject them at their face value. And let no one suppose that my position is 'holier than thou'. The book is addressed, in some embarrassment, to myself as much as to others.

The layout and content are designed both for the person who reads from cover to cover and for the one who dips. The subheadings are a guide. Each chapter stands largely on its own, but is linked to others. The summaries which start each chapter give an immediate overview of the book which I hope will entice the reader into the chapters rather than provide a substitute for reading them. I have also tried to think of the eight chapters as potential texts for seminars or lectures in educational and training institutions. On the subject of references, where a person is stated as a source of information but no reference is given, this usually indicates a personal communication.

Many people have, directly or indirectly, contributed to this book. Although I do not agree with all they say, the writings of Fritz Schumacher and Ivan Illich have influenced me. Without the enthusiasm and pressure of Milton Esman, John Montgomery and Peter Knight I might never have written the paper on 'Rural Poverty Unperceived: Problems and Remedies' which started all this off. Material from that paper is used with permission from the World Bank which published it as *Staff Working Paper* No. 400 and from *World Development* which published it in Vol. 9, No. 1, January 1981, pp. 1–19. Many others have helped. To colleagues at the Institute of Development Studies at the University of Sussex and in the Ford Foundation, New Delhi, I am indebted for comments, argument and insight over a long period, and for tolerance and support which have given me the opportunity and will to write. Others who have made extensive and valuable criticisms and comments for which I am grateful include Sarah Hamilton, John Hatch, Guy Hunter, Janice Jiggins, Narpat Jodha, Bruce Johnston, James Leach, David Nabarro, Arnold Pacey, Nigel Padfield, Ingrid Palmer, Anil Shah, Rupert Sheldrake, K. K. Singh, and Douglas Thornton. Four people have made a special contribution. Richard Jolly enabled me to say no to other demands on my time, and without him this book would not have been written; and Anthony Bottrall, Charles Elliott and John Harriss all read through the first draft, saw it as a whole, and made constructive criticisms about content, organisation and detail which led to a major revision.

There are many others, not named, who have assisted with information and comment. To list them all would be impossible and to list a few invidious. I ask them to accept my thanks offered generally, for they know who they are. Having said all that, my greatest debt is to Jennifer, my wife. She reviewed the draft and helped me to improve what the book says; but much more important, she has been a continuous source of insights and ideas which have enabled me to see, feel and think differently. In a sense, this book is the fruit of a collective effort by the few I have named and the many I have not named. Nevertheless, I expect all of them will find parts with which they disagree, and the responsibility for errors of fact and judgement is mine alone.

Finally, let me thank those who have typed the book at different stages: Susan Saunders and Ellen Miller at the Institute of Development Studies, University of Sussex, and I. D. Khurana at the Ford Foundation, New Delhi. Their cheerfulness, careful work, patience and inventiveness with obscure manuscripts made writing easier, quicker and more of a pleasure than I deserved.

When William Jansen was interviewing poor people in rural Bangladesh, one asked him: 'Gentleman, whatever are you writing so much about the poor people? God, himself, does not love the poor people: so what help will your writing do?' That is a discomforting question which I cannot answer. So let me pass it on now to the reader.

For those who are last
and those who put them first

We are grateful to Teaching Aids
at Low Cost (TALC) for assistance
in the production of this book.

TALC AND THE TROPICAL CHILD HEALTH UNIT
OF THE INSTITUTE OF CHILD HEALTH, LONDON,
RECEIVED ASSISTANCE IN THE PRODUCTION OF
THIS BOOK AS A LOW COST EDITION FROM THE
Swedish International Development Authority

CHAPTER ONE
Rural poverty unperceived

The past quarter century has been a period of unprecedented change and progress in the developing world. And yet despite this impressive record, some 800 million individuals continue to be trapped in what I have termed absolute poverty: a condition of life so characterized by malnutrition, illiteracy, disease, squalid surroundings, high infant mortality, and low life expectancy as to be beneath any reasonable definition of human decency.

Robert S. McNamara, 1978, Foreword to *World Development Report*

The latter-day ease with which officials, businessmen and international 'experts' can speed along highways has the effect of almost entirely divorcing the rural sector from its urban counterpart; no longer must 'dry season' earth roads be negotiated and preparations made for many night stops in villages. It is now possible to travel 100 kilometres or more through the countryside (in limousine comfort) during the morning, and get back to the city in time for lunch having observed nothing of the rural condition. Why the hurry (one may ask)? Can there be no digression from the superhighway path? But account has to be taken of the age-old motive for human action, *fear of involvement*. Time does not change human behaviour so very much, and still today what might be unpleasant or personally demanding, but is not actually seen, is often ignored.

Margaret Haswell, 1975, in *The Nature of Poverty*, pp. 213–14

What the eye does not see, the heart does not grieve about.
Old English Proverb

Outsiders are people concerned with rural development who are themselves neither rural nor poor. Many are headquarters and field staff of government organisations in the Third World. They also include academic researchers, aid agency personnel, bankers, businessmen, consultants, doctors, engineers, journalists, lawyers, politicians, priests, school teachers, staff of training institutes, workers in voluntary agencies, and other professionals. Outsiders underperceive rural poverty. They are attracted to and trapped in urban 'cores' which generate and communicate their own sort of knowledge while rural 'peripheries' are isolated and neglected. The direct rural experience of most urban-based outsiders is limited to the brief and hurried visits, from urban centres, of rural development tourism. These exhibit six biases against contact with and learning from the poorer people. These are *spatial* – urban, tarmac and roadside; *project* – towards places where there are projects; *person* – towards those who are better off, men rather than women, users of services and adopters of practices rather than non-users and non-adopters, and those who are active, present and living; *seasonal*, avoiding the bad times of the wet season; *diplomatic*, not seeking out the poor for fear of giving offence; and *professional*, confined to the concerns of the outsider's specialisation. As a result, the poorer rural people are little seen and even less is the nature of their poverty understood.

We, the outsiders

The extremes of rural poverty in the third world are an outrage. Faced with the facts, few would disagree with that statement. The outrage is not just that avoidable deprivation, suffering and death are intolerable; it is also that these coexist with affluence. Most of those who read this book will, like the writer, be immeasurably better off than the hundreds of millions of poorer rural people, living in this same world, who have to struggle to find enough to eat, who are defenceless against disease, who expect some of their children to die. Whatever the estimates of numbers – and endless scholastic argument is possible about definitions, statistics and the scale and degree of deprivation – there are so many people who are so poor, the prospects of future misery are so appalling, and present efforts to eliminate that misery are so inadequate, that numbers are almost irrelevant in seeing what to do next. So much needs to be done, so much more radically, that no estimates, however optimistic, could undermine the case for trying to do much more, much better, and faster.

But who should act? The poorer rural people, it is said, must help themselves; but this, trapped as they are, they often cannot do. The initiative, in enabling them better to help themselves, lies with outsiders who have more power and resources and most of whom are neither rural nor poor. This book has been written in the hope that it will be of some use to these outsiders, especially but not only those directly engaged in rural development work. Many of these are headquarters, regional, district and subdistrict staff in government departments in Third World countries – in administration, agriculture, animal husbandry, community development, cooperatives, education, forestry, health, irrigation, land development, local government, public works, water development, and the like. They also include all others, from and in both rich and poor countries, whose choices, action and inaction impinge on rural conditions and the poorer rural people – including academic researchers, aid agency and technical cooperation personnel, bankers, businessmen, consultants, doctors, engineers, journalists, lawyers, politicians, priests, school teachers, staff of training institutes and voluntary agencies, and other professionals.

We, these outsiders, have much in common. We are relatively well-off, literate, and mostly urban-based. Our children go to good schools. We carry no parasites, expect long life, and eat more than we need. We have been trained and educated. We read books and buy newspapers. People like us live in all countries of the world, belong to all nationalities, and work in all disciplines and professions. We are a class.

The puzzle is that we, the people of this class, do not do more. If any of us had a sick or starving child in the room with us, we imagine we would do something about it. A child crying from pain or hunger in a room is hard to shut out; it pins responsibility onto those present and demands, impels, action. Yet we live in a world where millions of children cry from avoidable hunger and pain every day, where we can do something about it, and where for the most part we do little. There are some exceptions: they include those who live with, work with, and learn with the poorer rural people. A very few have chosen to reject the privileges of our class for themselves and their children and live lives which reflect their convictions. Yet most of us manage to evade those choices. What is the difference between the room and the world? Why do we do so much less than we could?

There are many explanations. One is the simple fact of distance. The child is not in the room with us, but in Bihar, Bangladesh, the Sahel or a nameless camp for refugees, out of sight, sound and mind. Time, energy, money, imagination and compassion are finite. People deal first with what confronts them.

Rural poverty is remote. It is even remote, most of the time, for those outsiders who are 'working in the field' but who are urban-based, like government and other professionals in regional, district or subdistrict headquarters. One starting point is then to examine, as this first chapter does, the nature of contact or lack of contact between urban-based outsiders concerned with rural development and the poorer rural people.

But distance is only the easy part of the explanation. There is also a wilful element of choice. Outsiders choose what to do – where to go, what to see, and whom to meet. What is perceived depends on the perceiver. Outsiders have their own interests, preferences and preconceptions, their own rationalisations, their own defences for excluding or explaining the discordant and the distressing. Selfishness is a powerful force. Putting one's family first seems natural and good, and 'charity begins at home' is a great let-out. Disillusion with development failures and a knowing cynicism about 'where the money goes' are given as reasons for doing nothing. Outsiders are often ignorant about rural poverty but do not want to know what they do not know. The less they have of direct and discordant contact and learning, and the less they know, so the easier it is for myth to mask reality. Outsiders as a class need comforting beliefs: that rural deprivation is not so bad; that their prosperity is not based on it; that the poorer people are used to it and like life their way; or that they are lazy and improvident and have brought it on themselves. Such convenient beliefs about social ills are suspect. They present a challenge for analysis. Critical self-examination is not easy. What follows in this book is only one beginning. It invites outsiders concerned with rural development to analyse the ways they learn, think, feel and act, and to see how these might be changed to make things less bad especially for the more deprived of those who are rural and poor.

Cores and peripheries of knowledge

The argument is set in a context of cores and peripheries of knowledge. Globally, these reflect a gradient from extremes of wealth to extremes of poverty. At one end there coexist rich, urban, industrialised, high status cores, and at the other, poor, rural, agricultural and low status peripheries. In the cores there is a mutual attraction and reinforcement of power, prestige, resources, professionals, professional training and the capacity to generate and disseminate knowledge. Both internationally and within individual third world countries, centripetal forces draw resources and educated people away from the peripheries and in towards the cores. Within third world countries, skills migrate from rural to urban areas, and from smaller to larger urban centres, feeding in turn the international flows of the brain drain. The centripetal system is self-reinforcing. Staff and resources drawn to the rich, urban, industrial cores add to the mass which generates prestige and rewards and attracts yet more staff and resources.

At the very centre are the black holes of professionalism – space programmes and defence in the United States and the USSR – sucking in staff and resources which are then lost to sight. Research points where the rich and powerful direct it – to arms, rockets, chips, cars, minerals, chemicals, diseases of the affluent and ageing, and the mechanised agriculture of temperate climates. Trained and drawn towards these cores are those professions which directly touch rural life, like medicine, engineering and agriculture, looking inwards to the establishments of the rich world as their arbiters of priorities and excellence. At the other extreme, there are government staff, voluntary workers, and researchers pushing out into, clinging to, or stuck in the rural periphery. Some have failed to move inwards into the system; some have rejected it; and some are trying to change it.

The allocation of resources to research is a measure of the imbalances of the system. Overwhelmingly, research and development (R and D) expenditure in the world is concentrated in the industrialised countries. It might have been thought that rural poverty deserved a higher priority than defence, yet we find over 50 per cent of the research scientists in the world engaged in defence work (DF 1979). In 1980, although there was a stockpile of nuclear weapons with one million times the destructive power of the Hiroshima bomb, the USA and USSR were spending well over $100 million *per day* on upgrading their nuclear arsenals, compared with a recent figure of a derisory $60 million *per year* devoted to tropical diseases research (Sivard 1980, p. 5; Walsh and Warren 1979, p. 20). Over a wide range, there remains deep ignorance about many researchable physical and social aspects of rural life – soil erosion, the diarrhoeas, the political economy of pastoralism, drudgery-reducing technology for rural women, the management of canal irrigation systems, levels of human calorie requirements, seasonal interactions between nutrition work, sickness and indebtedness, the relative importance of different contingencies which make poor people poorer, forms of organisation to overcome the tragedy of the commons – to make but a quick, short list which could be extended many times over.

A tiny diversion of resources to increase sensitive research on topics such as these might mitigate the misery of many millions of people. But the mainstream of the R and D system flows in another another direction, passing by on the other side, and drawing resources away after it.

Why this should be so can be understood partly in terms of how education and professional training articulate with the preoccupations, power and prestige of the cores. Prolonged professional conditioning has built biases of perception deep into many of those concerned with rural development. These direct attention towards whatever is urban, industrial, 'high' technology, capital-intensive, appropriate for temperate climates, and marketed and exported; to the neglect of what is rural, agricultural, 'low' technology, labour-intensive, appropriate for tropical climates, retained by the household and locally consumed. Many interlocking influences shape ambitions, mould ways of seeing things and sway choices of where in the world one is to work. These include textbooks, curricula, examination questions, professional journals, academic awards, national and international distinctions, professional values and ideas of sophistication, the media, the priority accorded to armaments and security, the desire of elites for international mobility. Most professionals face away from rural areas; most live in towns. And even among that minority who face the other way, or who live in rural areas, their conditioning has often disabled them. They direct their attention to those with whom they have most in common – the less poor rural people. They see and link in with whatever they can find which is familiar and prestigious – with whatever is modern, marketed, urban in origin, and sophisticated. They prescribe for only that specialised part of the diverse rural reality for which their training has prepared them.

At its ugliest, such professional training inculcates an arrogance in which superior knowledge and superior status are assumed. Professionals then see the rural poor as ignorant, backward and primitive, and as people who have only themselves to blame for their poverty. Social Darwinism then lives again in the ideologies of the prosperous and therefore virtuous elites looking out on the rural mass, the poverty of whose members reflects their lack of virtue. The very phrase – 'the rural mass' – fosters stereotypes, convenient glosses hiding ignorance of the reality. Not only do urban-based professionals and officials often not know the rural reality; worse, they do not know that they do not know.

The urban trap

It is by no means only the international system of knowledge and prestige, with its rewards and incentives, that draws professionals away from rural areas and up through the hierarchy of urban and international centres. They are also attracted and held fast by better houses, hospitals, schools, communications, consumer goods, recreation, social services, facilities for work, salaries and career prospects. In third world countries as elsewhere, academics, bureaucrats, foreigners and journalists are all drawn to towns or based in them. All are victims, though usually willing victims, of the urban trap. Let us consider them in turn.

For academics, it is cheaper, safer and more cost-effective in terms of academic output, to do urban rather than rural research. If rural work is to be done, then peri-urban is preferable to work in remoter areas. Rural research is carried out mainly by the young and inexperienced. For them rural fieldwork is a rite of passage, an initiation which earns them the right to do no more, giving them a ticket to stay in the town. But the fieldwork must first be performed in the correct manner as prescribed by custom. The social anthropologist has to spend a year or so in the village, the sociologist to prepare, apply, analyse and write up a questionnaire survey. The ritual successfully completed, the researcher is appointed and promoted. Marriage and children follow. For women, pregnancy and child care may then dislocate a career and prevent further rural exposure. For men, family responsibilities tie less, but still restrain. Promotion means responsibility and time taken with teaching, supervising, administration, and university or institutional politics. The stage of the domestic cycle with small children means accumulation of responsibilities – driving children to school and picking them up again, family occasions, careful financial management to make ends meet, moonlighting and consultancies to supplement a meagre salary – all of which take time.

The researcher has now learnt enough to make a contribution to rural research. He or she has the confidence and wit to explore new ideas and to pursue the unexpected. There is evidence enough of this in the books by social anthropologists who have undertaken second and subsequent spells of fieldwork. But it is precisely at this time that the able academic is chained to desk, lectern and home. If the university rewards ability, then the more able persons are likely to be most trapped. Ageing, ability, promotion and the domestic cycle conspire to prevent further rural contact.

The amalgam which glues these forces together and finally

immobilises the would-be rural researcher in mid-career is overcommitment. It is a mystery why so many of the presumably intelligent people who do research are so miserably incompetent at managing their own lives. Academics can be found who are simultaneously supervising half a dozen theses (if their students can get near them), managing a major research project (actually managed by a junior administrator and by field staff), lecturing (from old notes or off-the-cuff), sitting on a dozen committees (or sending in, or failing to send in, apologies for absence), writing a couple of books (or adding notes to the draft by the junior author), developing a new curriculum or course (which for lack of time ends up much like a previous one), and carrying out a consultancy for an aid agency (which, for inescapable financial reasons, takes priority over all else). To judge from a limited and scattered sample, I suspect a positive correlation between overcommitment at work and size of family, though whether this reflects a lack of restraint and planning in both domains may be an idle speculation. But for such people, overcommitment is an addiction. In extreme cases, they take on more and more and complete less and less, complete it less and less well and, as they become more eminent, are less and less likely to be told their work is bad. Needless to say, there is also less and less time for any direct rural exposure: for the demands of students, researchers, administrators, committees, new curricula, books and consultancies all require presence in town. Ambition, inefficiency, and an inability to say no, tie the academic down, as an urban prisoner. Parole is rare and brief: rural contact is restricted to hectic excursions from the urban centre where the university or institute is sited.

For government staff, there are similar pressures and patterns. On first appointment, when ignorant and inexperienced, technical or administrative officers are posted to the poorer, remoter, and politically less significant areas. Those who are less able, less noticed, or less influential, remain there longer. The more able, and those who come favourably to attention or who have friends in headquarters, are soon transferred to more accessible or more prosperous rural areas, or to urban centres. Administration is, anyway, an urban-based and urban-biased activity. So with promotion, contact with rural areas, especially the remoter ones, recedes. If a serious error is committed, or a powerful politician offended, the officer may earn a 'penal posting', to serve out punishment time in some place with poor facilities – a pastoral area, an area without irrigation, an area distant from the capital, an area which is hot and unhealthy – in short, a place where poorer people will be found. But the pull of urban life will remain: children's education, chances of promotion, congenial company, consumer goods, cinemas, libraries, hospitals, and quite simply power; all drawing bureaucrats away from rural areas and towards the major urban and administrative centres.

Once established in offices in the capital city, or in the regional or provincial headquarters, bureaucrats too are trapped. Unless they are idle and incompetent, or exceptionally able and well supported, they are quickly overcommitted. They are tied down by committees, subcommittees, memoranda, reports, urgent papers, personnel problems, financial management, and the professional substance of their work. There are political demands to which they must be able to react swiftly and efficiently. There are times of the year, during the budget cycle, when they cannot contemplate leaving their desks. The very emphasis on agricultural and rural development creates work, which holds them in their offices. If the government is inactive, they may be relatively free. But the more the government tries to do, so the more paperwork is generated, the more coordination and integration are called for, the more reports have to be written and read, and the more inter-ministerial and inter-departmental coordination and liaison committees are set up. The more important these committees become, so the more members they have, the longer their meetings take, and the longer their minutes grow. The demands of aid agencies are a final straw, requiring data, justifications, reports, evaluations, visits by missions, and meetings with ministers. More activity, more aid, more projects, more coordination – all these mean more time in the office and less in the field.

Foreigners are also urban-based and urban-biased. Foreigners in third world countries who are concerned with rural development and rural poverty include staff in voluntary agencies and aid organisations, technical cooperation personnel of various sorts, and consultants. Many voluntary agency workers and a few technical cooperation staff do live in rural areas. But most of these foreigners are also urban-based, many of them in capital cities, and have the familiar problems of paperwork, meetings and political and family pressures which tie them there. In addition their rural movements may be restricted by a suspicious government, or smothered in protocol. Their perceptions vary from the acute and correct to the naive and mistaken. They often labour under the notorious difficulties and distortions of having to rely on interpreters, of being taken on conducted tours, and of misleading responses from those met.

A final group, neglected yet vital for the formation of opinion

9

about rural life, are journalists. They combine the most direct access to mass media with the severest constraints on rural exposure. Journalists who wish to visit a rural area have three problems. First, they must persuade their editor that the visit is worthwhile. This is difficult. In terms of news, it is almost always quicker and cheaper to look for and write up an urban story; moreover a disproportion of newspaper readers are urban dwellers interested in urban news. Second, journalists must be sure to get a story. This usually means a visit either in special company (for example, the Prime Minister's visit to a region) with an official entourage and all that goes with it, or to an atypical rural place where there is either a project or a disaster. Third, journalists cannot hang around. They must find out what they want quickly and write it up quickly. Checking information is difficult, and with rural people who are unlikely to read what is written let alone sue, the incentive to check it is low. It is the one-off rushed and unconfirmed interview which appears in quotation marks in the newspaper article. Like academics, bureaucrats and foreigners, journalists are both actors and victims in the brief rural visit.

Rural development tourism

For all these urban-based professionals, the major source of direct experience of rural conditions is, then, rural development tourism, the phenomenon of the brief rural visit. This influences and is part of almost all other sources of information. It is extremely widespread, with perhaps tens of thousands of cases daily in third world countries. In spite of its prevalence, it has not to my knowledge been seriously analysed. This omission is astonishing until one reflects on the reasons. For academic analysis, rural development tourism is too dispersed and ephemeral for convenient rigour, not neatly in any disciplinary domain, and barely conceivable as the topic for a thesis. For practical professionals engaged in rural development, it is perhaps too near the end of the nose to be in focus. Rural development tourism is, moreover, a subject of anecdote and an object of shame. It generates stories for bar gossip rather than factors for comparative study, and evokes memories of personal follies one prefers not to expose to public ridicule. In any case, self-critical introspection is not one of the more prominent characteristics of rural developers. Yet it is through this rural development tourism, if at all, that 'core' (urban-based, professional, powerful) visitors see and meet those who are 'peripheral' (rural, uneducated, weak). The brief rural visits by 'core' personnel can scarcely fail to play a key part in forming their impressions and beliefs and influencing their decisions and actions.

Let us examine the phenomenon. The visits may be for one day or for several. The 'tourists' or visitors may come from a foreign country, a capital city, a seat of regional or provincial government, a district headquarters, or some smaller urban place. Most commonly they are government officials – administrators, health staff, agriculturalists, veterinarians, animal husbandry staff, educators, community developers, engineers, foresters, or inspectors of this and that; but they may also be private technical specialists, academic researchers, the staff of voluntary agencies, journalists, diplomats, politicians, consultants, or the staff of aid agencies. Differing widely in race, nationality, religion, profession, age, sex, language, interests, prejudices, conditioning and experience, these visitors nevertheless usually have three things in common: they come from urban areas; they want to find something out; and they are short of time.

Rural development tourism has many purposes and many styles. Technical specialists concerned with physical resources may in practice have little contact with rural people, and there may be little formality about their visits. Others – those concerned with administration and human development in its various forms – may in contrast be involved in many meetings with rural people. It is with these kinds of visits that we are primarily concerned. It is tempting to caricature, and exaggeration is built into any process of induction from anecdotes which are repeated and remembered because they make good stories. There are also differences between cultures, environments and individual tourists. But it may hold generally that the older, more senior, more important, and more involved with policy the tourist is, so the larger will be the urban centre from which he[1] leaves, and the more likely his visit is to be selective and formally structured. The more powerful professionals are, the less chance they have of informal learning.

A sketch can illustrate the problems[2] of such visits by the powerful, important, and distinguished. The visitor sets out late, delayed by last minute business, by colleagues, by subordinates or superiors anxious for decisions or actions before his departure, by a family crisis, by a cable or telephone call, by others taking part in the same visit, by mechanical or administrative problems with vehicles, by urban traffic jams, or by any one of a hundred forms of human error. Even if the way is not lost, there is enough fuel, and there are no breakdowns, the programme runs behind schedule. The visitor is encapsulated, first in a limousine, Landrover, Jeep or car and later in a moving entourage of officials and local notables – headmen, chairmen of village committees, village accountants, progressive farmers, traders, and the like.

Whatever their private feelings, (indifferent, suspicious, amused, anxious, irritated, or enthusiastic), the rural people put on their best face and receive the visitor well. According to ecology, economy and culture, he is given goats, garlands, coconut milk, coca-cola, coffee, tea or milk. Speeches are made. Schoolchildren sing or clap. Photographs are taken. Buildings, machines, construction works, new crops, exotic animals, the clinic, the school, the new road, are all inspected. A self-conscious group (the self-help committee, the women's handicraft class), dressed in their best clothes, are seen and spoken to. They nervously respond in ways which they hope will bring benefits and avoid penalties. There are tensions between the visitor's questions and curiosity, the officials' desire to select what is to be seen, and the mixed motives of different rural groups and individuals who have to live with the officials and with each other after the visitor has left. Time and an overloaded programme nevertheless are on the officials' side. As the day wears on and heats up, the visitor becomes less inquisitive, asks fewer questions, and is finally glad to retire, exhausted and bemused, to the circuit bungalow, the rest house, the guest house, the host official's residence, or back to an urban home or hotel. The village returns to normal, no longer wearing its special face. When darkness falls and people talk more freely, the visitor is not there.

Shortage of time, the importance of the visitor, and the desire for information separately or together influence what is perceived. Lack of time drives out the open-ended question; the visitor imposes meanings through what is asked. Checking is impossible, and prudent, hopeful, or otherwise self-serving lies become accepted as facts. Individually or in groups, people are neglected while formal actions and physical objects receive attention. Refugees in a rural camp in Tanzania said of UN and government officials that 'They come, and they sign the book, and they go', and 'They only talk with the buildings'. A villager in Senegal said to Adrian Adams concerning visitors: 'Ils ne savent pas qu'il y a ici des gens vivants'[3] (Adams, 1979, p. 477). Above all, on such visits, it is the poorer people who tend not to be seen, far less to be met.

Rural poverty unobserved: the six biases

Many biases impede outsiders' contact with rural poverty in general, and with the deepest poverty in particular. These apply not only to rural development tourists, but also to rural researchers and local-level staff who live and work in rural areas. Six sets of biases stand out:

i) Spatial biases: urban, tarmac and roadside

Most learning about rural conditions is mediated by vehicles. Starting and ending in urban centres, visits follow networks of roads. With rural development tourism, the hazards of dirt roads, the comfort of the visitor, the location of places to visit and places for spending the night, and shortages of both time and fuel dictate a preference for tarmac roads and for travel close to urban centres. The result is overlapping urban, tarmac and roadside biases.

Urban bias concentrates rural visits near towns and especially near capital cities and large administrative centres. But the regional distribution of the poorest rural people often shows a concentration in remoter areas – north-eastern Brazil, Zambia away from the line of rail, lower Ukambani in Kenya, the Tribal Districts of Central India, the hills of Nepal. In much of the developing world, some of the poorest people are being driven from those densely populated areas better served with communications and are being forced, in order to survive, to colonise less accessible areas, especially the savannahs and forests. Hard to reach from the urban centres, they remain largely unseen.

Tarmac and roadside biases also direct attention towards those who are less poor and away from those who are poorer. Visible development follows main roads. Factories, offices, shops and official markets all tend to be at the sides of main roads. Even agricultural development has a roadside bias: in Tamil Nadu agricultural demonstrations of new seeds and fertilisers have often been sited beside main roads; and on irrigation systems, roads follow canals so that the farms seen are those of the topenders who receive more water and not those of the tailenders who receive less or none. Services along roadsides are also better. An improved tarmac or all-weather surface can bring buses, electricity, telephone, piped water supply, and better access to markets, health facilities and schools. Services near main roads are better staffed and equipped: Edward Henevald found that two schools near a main highway in Sumatra had more than their

13

quota of teachers, while a school one kilometre off the road had less than its quota.

When roads are built, land values rise and those who are wealthier and more influential often move in if they can. In Liberia, new rural roads were followed by speculators rushing to acquire deeds and to buy or to displace local farmers (Cobb et al. 1980, pp. 12–16). For part of Western Kenya, Joseph Ssennyonga had described a similar tendency for the wealthier and more influential to buy up roadside plots, creating an 'elite roadside ecology' (1976: p. 9). So the poorer people shift away out of sight. The visitor then sees those who are better-off and their houses, gardens, and services, and not those who are poorer and theirs. Ribbon development along roadsides gives a false impression in many countries. The better the road, the nearer the urban centre, and the heavier the traffic, so the more pronounced is the roadside development and the more likely visitors are to see it and be misled.

Nor does spatial bias apply only to main roads. Within villages, the poorer people may be hidden from the main streets and the places where people meet. M. P. Moore and G. Wickremesinghe, reporting on a study of three villages in the Low Country of Sri Lanka, have this to say about 'hidden poverty':

> In retrospect at least, one of the most obvious aspects of poverty in the study villages is the extent to which it is concealed from view ... the proportion of 'poor' households ... varies from 14 per cent in Wattegama to 41 per cent in Weligalagoda. Yet one could drive along all the motorable roads in the villages and scarcely see a single 'poor' house. Here, as in most of rural Sri Lanka, wealthier households use their social and economic power to obtain roadside homestead sites. Not only do these confer easier access to such tangible services as buses, electricity connections or hawkers, but they provide such intangible benefits as better information and gossip from passers-by. Equally, the roadside dweller has a potential site for opening a small shop, especially if located near the all-important road junctions, which provide the focus of commercial and social life in almost all rural areas. To even see the houses of the poor one often has to leave the road. Many visitors, including public officers, appear not to do so very often.

> (1980, p. 59; emphasis added)

The same can be said of Harijan colonies in or near villages in

South India, and of Basarwa (Bushmen) in or outside the villages of the Kalahari. Peripheral residence is almost universal with the rural poor.

It is not just the movements of officials that are guided by these spatial biases of rural development tourism. Social science researchers are far from immune. There are honourable exceptions, but urban and tarmac biases are sometimes evident in choices of villages to study. Of all specialists, social anthropologists are perhaps the least susceptible, but even they sometimes succumb: as they have grown, Bangalore and Bangkok have each swallowed up a social anthropologist's village.[4] Again, when Indian institutions were urged to adopt villages, two research and training organisations in Bangalore, unknown to each other, included the same village: it can scarcely be a coincidence that it was close to the main Bangalore-Mysore road, a decent but convenient distance from Bangalore itself. Within villages, too, the central, more prosperous, core is likely to attract researchers.

Moore, again describing three villages in Sri Lanka, writes:

> Apart from the roadside issue, the core can exercise a great pull on the outsider who decides to do a few days' or a week's fieldwork. Apart from the facilities and the sense of being at the strategic hub of local affairs, it can claim a sense of history and tradition, to which sociologists especially appear vulnerable.
>
> (1981, p. 48)

He considers that sociologists writing on Sri Lanka have mostly focussed on core areas and completely ignored the peripheries. One may speculate about how generally the location of good informants and of facilities at the cores of villages prevent perception by social scientists of the peripheral poor.

Urban bias is further accentuated by fuel shortages and costs. When fuel costs rise dramatically, as they have done in recent years, the effect is especially marked in those poor countries which are without oil and also short of foreign exchange. The recurrent budgets of government departments are cut. Staff are difficult to shed, so the cuts fall disproportionately on other items. Transport votes are a favourite. Rural visits, research and projects shrink back from more distant, often poorer areas to those which are closer, more prosperous, and cheaper to visit.[5]

In Zambia, the travel votes of the Ministry of Agriculture and Water Development could buy in 1980 only one fifth of the petrol they could buy in 1973 (ILO, 1981, p. 74) and senior agricultural extension staff were virtually office-bound. In Bangladesh, similarly, district agricultural officers have been severely restricted in their use of vehicles. In India, cuts have occurred in transport allocations for staff responsible for supervising canal irrigation: the likely effects include less supervision leading to less water reaching the already deprived areas and less staff awareness of what is happening there. Every rise in oil prices impoverishes the remoter, poorer people by tilting the urban-rural terms of trade against them, and at the same time reduces the chances of that deprivation being known. Visits, attention and projects are concentrated more and more on the more accessible and more favoured areas near towns.

ii) Project bias

Rural development tourism and rural research have a project bias. Those concerned with rural development and with rural research become linked to networks of urban-rural contacts. They are then pointed to those rural places where it is known that something is being done – where money is being spent, staff are stationed, a project is in hand. Ministries, departments, district staff, and voluntary agencies all pay special attention to projects and channel visitors towards them. Contact and learning are then with tiny atypical islands of activity which attract repeated and mutually reinforcing attention.

Project bias is most marked with the showpiece: the nicely groomed pet project or model village, specially staffed and supported, with well briefed members who know what to say and which is sited a reasonable but not excessive distance from the urban headquarters.[6] Governments in capital cities need such projects for foreign visitors; district and subdistrict staff need them too, for visits by their senior officers. Such projects provide a quick and simple reflex to solve the problem of what to do with visitors or senior staff on inspection. Once again, they direct attention away from the poorer people.

The better known cases concern those rural development projects which have attracted international attention. Any roll of honour would include the Anand Dairy Cooperatives in India; the Chilalo Agricultural Development Unit in Ethiopia; the Comilla Project in Bangladesh; the Gezira Scheme in Sudan; the Intensive Agricultural Districts Programme (IADP) in India; Lilongwe in Malaŵi; the Muda Irrigation Project in Malaysia; the Mwea Irrigation Settlement in Kenya; and some *ujamaa* villages in Tanzania. These have been much visited and much studied.

Students seeking doctorates have read about them and then sought to do their fieldwork on them.[7]

Research generates more research; and investment by donors draws research after it and funds it. In India, the IADP, a programme designed to increase production sharply in a few districts which were well endowed with water, exercised a powerful attraction to research compared to the rest of India. An analysis (Harriss, 1977, pp. 30–34) of rural social science research published in the Bombay *Economic and Political Weekly* showed an astonishing concentration in IADP districts, and an almost total neglect of the very poor areas of central India. In a different way, the Comilla Project may also have misled, since Comilla District has the lowest proportion of landless of any district in Bangladesh. Research on *ujamaa* in Tanzania in the clusters of villages (the Ruvuma Development Association, Mbambara, and Upper Kitete) which were among the very few in the whole country with substantial communal agricultural production, sustained the myth that such production was widespread. Research, reports and publications have given all these atypical projects high profiles, and these in turn have generated more interest, more visitors, and yet more research, reports and publications.

Fame forces project managers into public relations. More and more of their time has to be spent showing visitors around. Inundated by the celebrated, the curious, and the crass – prime ministers, graduate students, women's clubs, farmers' groups, aid missions, evaluation teams, school parties, committees and directors of this and that – managers set up public relations units and develop a public relations style. Visitors then get the treatment. A fluent guide follows a standard route and a standard routine. The same people are met, the same buildings entered,[8] the same books signed, the same polite praise inscribed in the book against the visitors' names. Questions are drowned in statistics; doubts inhibited by handouts. Inquisitive visitors depart loaded with research papers, technical evaluations, and annual reports which they will probably never read. They leave with a sense of guilt at the unworthy scepticism which promoted their probing questions, with memories of some of those who are better-off in the special project, and impressed by the charisma of the exceptional leader or manager who has created it. They write their journey reports, evaluations and articles on the basis of these impressions.

For their part, the project staff have reinforced through repetition the beliefs which sustain their morale; and their projects take off into self-sustaining myth. But in the myth is the

17

seed of tragedy, as projects are driven down this path which leads, step-by-step to self-deception, pride, defensiveness, and ultimately debunking.

iii) Person biases

The persons with whom rural development tourists, local-level officials, and rural researchers have contact, and from whom they obtain impressions and information, are biased against poorer people.

a) *Elite bias*   'Elite' is used here to describe those rural people who are less poor and more influential. They typically include progressive farmers, village leaders, headmen, traders, religious leaders, teachers, and paraprofessionals. They are the main sources of information for rural development tourists, for local-level officials, and even for rural researchers. They are the most fluent informants. It is they who receive and speak to the visitors; they who articulate 'the village's' interests and wishes; their concerns which emerge as 'the village's' priorities for development. It is they who entertain visitors, generously providing the expected beast or beverage. It is they who receive the lion's share of attention, advice and services from agricultural extension staff (Chambers, 1974, p. 58; Leonard, 1977, Ch. 9). It is they who show visitors the progressive practices in their fields. It is they too, who, at least at first, monopolise the time and attention of the visitor.

Conversely, the poor do not speak up. With those of higher status, they may even decline to sit down. Weak, powerless and isolated, they are often reluctant to push themselves forward. In Paul Devitt's words:

> The poor are often inconspicuous, inarticulate and unorganised. Their voices may not be heard at public meetings in communities where it is customary for only the big men to put their views. It is rare to find a body or institution that adequately represents the poor in a certain community or area. Outsiders and government officials invariably find it more profitable and congenial to converse with local influentials than with the uncommunicative poor.
> (1977, p. 23)

The poor are a residual, the last in the line, the most difficult to find, and the hardest to learn from: 'Unless paupers and poverty are deliberately and persistently sought, they tend to remain effectively screened from outside inquirers' (*ibid.*, p. 24).

b) *Male bias*   Most local-level government staff, researchers and other rural visitors are men. Most rural people with whom they establish contact are men. Female farmers are neglected by male agricultural extension workers. In most societies women have inferior status and are subordinate to men. There are variations and exceptions, but quite often women are shy of speaking to male visitors. And yet poor rural women are a poor and deprived class within a class. They often work very long hours, and they are usually paid less than men. Rural single women, female heads of households, and widows include many of the most wretched and unseen people in the world.

c) *User and adopter biases*   Where visits are concerned with facilities or innovations, the users of services and the adopters of new practices are more likely to be seen than are non-users and non-adopters. This bias applies to visitors who have a professional interest in, say, education, health or agriculture, to local-level officials, and to researchers. They tend to visit buildings and places where activity is concentrated, easily visible, and hence easy to study. Children in school are more likely to be seen and questioned than children who are not in school; those who use the health clinic more than those who are too sick, too poor, or too distant to use it; those who come to market because they have goods to sell or money with which to buy, more than those who stay at home because they have neither; members of the cooperative more than those who are too poor or powerless to join it; those who have adopted new agricultural, health or family planning practices more than those who have not.

d) *Active, present and living biases*   Those who are active are more visible than those who are not. Fit, happy, children gather round the Jeep or Land Rover, not those who are apathetic, weak and miserable. Dead children are rarely seen. The sick lie in their huts. Inactive old people are often out of sight; a social anthropologist has recorded how he spent some time camping outside a village in Uganda before he realised that old people were starving (Turnbull, 1973, p. 102). Those who are absent or dead cannot be met, but those who have migrated and those who have died include many of the most deprived. Much of the worst poverty is hidden by its removal.

iv) *Dry season biases*

Most of the poor rural people in the world live in areas of marked
wet-dry tropical seasons. For the majority whose livelihoods
depend on cultivation the most difficult time of the year is usually
the wet season, especially before the first harvest. Food is short,
food prices are high, work is hard, and infections are prevalent.
Malnutrition, morbidity and mortality all rise, while body
weights decline. The poorer people, women and children are
particularly vulnerable. Birth weights drop and infant mortality
rises. Child care is inadequate. Desperate people get indebted.
This is both the hungry season and the sick season. It is also the
season of poverty ratchet effects, that is, of irreversible downward
movements into poverty through the sale or mortgaging of assets,
the time when poor people are most likely to become poorer.[9]

The wet season is also the unseen season. Rural visits by the
urban-based have their own seasonality.

> Nutritionists take care to plan
> to do their surveys when they can
> be sure the weather's fine and dry,
> the harvest in, food intake high.
>
> Then students seeking Ph.D.s
> believe that everyone agrees
> that rains don't do for rural study
> —suits get wet and shoes get muddy
>
> And bureaucrats, that urban type,
> wait prudently till crops are ripe,
> before they venture to the field
> to put their question: 'What's the yield?'

For monsoonal Asia, which has its major crop towards the end of
the calendar year, it is also relevant that:

> The international experts' flights
> have other seasons; winter nights
> In London, Washington and Rome
> are what drive them, in flocks, from home

since they then descend on India and other countries north of the
equator in January and February at precisely the time of least
poverty and when marriages and celebrations are to be seen and
heard.

Some opposite tendencies, however, deserve to be noted:

> And northern academics too
> are seasonal in their global view
> For they are found in third world nations
> mainly during long vacations.

North of the equator this means visits at the bad time of the monsoon in much of Asia and of the rains of West Africa. There are also professionals like agriculturalists and epidemiologists whose work demands rural travel during the rains, for that is when crops grow and bugs and bacteria breed.

But the disincentives and difficulties are strong. The rains are a bad time for rural travel because of the inconveniences or worse of floods, mud, landslides, broken bridges; and getting stuck, damaging vehicles, losing time, and enduring discomfort. In some places roads are officially closed. In the South Sudan there is a period of about two months after the onset of the rains when roads are impassable but when there is not yet enough water in the rivers for travel by boat. Many rural areas, especially those which are remote and poor, are quite simply inaccessible by vehicle during the rains. The worst times of the year for the poorer people are thus those that are the least perceived by urban-based outsiders.

Once the rains are over such visitors can however travel more freely. It is in the dry season, when disease is diminishing, the harvest in, food stocks adequate, body weights rising, ceremonies in full swing, and people at their least deprived, that there is most contact between urban-based professionals and the rural poor. Not just rural development tourism, but rural appraisal generally is susceptible to a dry season bias. A manual for assessing rural needs warns of an experience when 'Once, the jeeps needed for transporting the interviewers were recalled for a month *during the few precious months of the dry season*' (Ashe, 1979, p. 26; my emphasis). Whole institutes concentrate their field research in the dry seasons; the rains are for data analysis and writing up with a good roof over one's head. Concern to avoid inconveniencing respondents when they are busy and exhausted with agricultural activities provides a neat justification, both practical and moral, for avoiding research during the rains. Many factors thus conspire to ensure that the poorest people are most seen at precisely those times when they are least deprived; and least seen when things are at their worst.

v) *Diplomatic biases: politeness and timidity*

Urban-based visitors are often deterred by combinations of politeness and timidity from approaching, meeting, and listening to and learning from the poorer people. Poverty in any country can be a subject of indifference or shame, something to be shut out, something polluting, something, in the psychological sense, to be repressed. If honestly confronted, it can also be profoundly disturbing. Those who make contact with it may offend those who are influential. The notables who generously offer hospitality to the visitor may not welcome or may be thought not to welcome, searching questions about the poorer people. Senior officials visiting junior officials may not wish to examine or expose failures of programmes intended to benefit the poor. Politeness and prudence variously inhibit the awkward question, the walk into the poorer quarter of the village, the discussion with the working women, the interviews with Harijans. Courtesy and cowardice combine to keep tourists and the poorest apart.

vi) *Professional biases*

Finally, professional training, values and interests present problems. Sometimes they focus attention on the less poor: agricultural extension staff trained to advise on cash crops or to prepare farm plans are drawn to the more 'progressive' farmers; historians, sociologists and administrators, especially when short of time, can best satisfy their interests and curiosity through informants among the better-educated or less poor; those engaged in family welfare and family planning work find that bases for the adoption of any new practices can most readily be established with better-off, better-educated families. But sometimes, in addition, professional training, values and interests do focus attention directly on the poor. This is especially so in the fields of nutrition and health, where those wishing to examine and to work with pathological conditions will tend to be drawn to those who are poorer.

More generally, specialisation, for all its advantages, makes it hard for observers to understand the linkages of deprivation. Rural deprivation is a web in which poverty (lack of assets, inadequate stocks and flows of food and income), physical weakness and sickness, isolation, vulnerability to contingencies, and powerlessness all mesh and interlock.[10] But professionals are trained to look for and see much less. They are programmed by their education and experience to examine what shows up in a bright but slender beam which blinds them to what lies outside it.

Knowing what they want to know, and short of time to find it out, professionals in rural areas become even more narrowly single-minded. They do their own thing and only their own thing. They look for and find what fits their ideas. There is neither inclination nor time for the open-ended question or for other ways of perceiving people, events and things. 'He that seeketh, findeth.' Visiting the same village, a hydrologist enquires about the water table, a soils scientist examines soil fertility, an agronomist investigates yields, an economist asks about wages and prices, a sociologist looks into patron-client relations, an administrator examines the tax collection record, a doctor investigates hygiene and health, a nutritionist studies diets, and a family planner tries to find out about attitudes to numbers of children. Some of these visiting professionals may be sensitive to the integrated nature of deprivation, but none is likely to fit all the pieces together, nor to be aware of all the negative factors affecting poorer people.

Specialisation prevents the case study which sees life from the point of view of the rural poor themselves; but where such case studies are written (e.g. Gulati, 1981; Howes, 1980; Ledesma, 1977; Lewis, 1959) their broader spread helps understanding and points to interventions which specialists miss. In contrast, narrow professionalism of whatever persuasion leads to diagnoses and prescriptions which underestimate deprivation by recognising and confronting only a part of the problem.

The unseen and the unknown

The argument must not be overstated. To all of these biases, exceptions can be found. There are government programmes, voluntary organisations, and research projects that seek out those who are more remote and poorer. Some projects and programmes, such as those for the weaker sections and vulnerable classes in rural India, have an anti-poverty focus. Person biases can work the other way: women's groups and women's programmes attract attention; doctors see those who are sick; nutritionists concentrate on the malnourished; agriculturalists and epidemiologists alike may have professional reasons for travel during the rains; and during an agricultural season, a daytime visit to a village may provide encounters with the sick, aged and very young, and not with the able-bodied who are out in the fields. Such exceptions must be noted. At the same time, there are dangers of underestimating the force of the biases by failing to see how they interlock and by underestimating their incidence.

The way in which spatial, project, person, dry season, politeness/timidity and professional biases interact can be seen by analysing almost any example of an urban-based outsider investigating rural conditions. With many 'insights' and beliefs about rural life, the several biases can and do reinforce each other. The prosperity after harvest of a male farmer on a project beside a main road close to a capital city may colour the perceptions of a succession of officials and dignitaries. The plight of a poor widow starving and sick in the wet season in a remote and inaccessible area may never in any way impinge on the consciousness of anyone outside her own community.

Nor are those professionals and rural staff who originate from rural areas, who have a home, second home, or farm there, or who live and work there, immune from these tendencies. Three examples can illustrate that their perceptions too can be powerfully distorted by the biases.

The first example is from a densely populated part of western Kenya. Junior agricultural extension staff and home economics workers were each given a random sample of 100 households to survey. The households were in the area where they worked. After the survey, those who had conducted it all considered that the sample had been unfairly weighted against the more progressive and better educated households, over-representing those that were poorer. One of the agricultural staff complained that of his 100 households, only one had an exotic grade cow, and that there would have been several more if the sample had been truly representative. In reality, however, in that area there was only one exotic grade cow for every 200 households, so each sample of 100 had only a 50 per cent chance of including one at all. A home economics worker said that she was appalled at the poverty she had encountered among her sample. On two occasions she had burst into tears at what she had found. She had not known that there was such misery in the area. 'These people,' she said, 'do not come to my meetings.'

For the same area, David Leonard (1977, p. 178) has documented the marked tendency for extension staff to visit progressive farmers, and not to visit non-innovators (57 per cent of visits to the 10 per cent who were progressives and only 6 per cent to the 47 per cent who were non-innovators). Thus, it is not only outsiders who are affected by anti-poverty biases. Local-level rural staff are also affected, and unless there are strong countervailing incentives, they too will underperceive deprivation in the very areas where they work.

The second example is from a study by Moore and Wickremesinghe (forthcoming, p. 98) in Sri Lanka. After observing how the houses of the poor are physically hidden from the core of the villages they studied, and how public officers appear not to see them very often, Moore and Wickremesinghe noted:

> Although most of the rural population ... are poor and dependent in part or whole on wage labour, one hears comments of the nature: 'Of course, most of the people around here have some job or little business in Colombo'.

The implication of such comments was that most people in the villages had other incomes and a modest well-being. This might be true of those who lived at the centres of the villages, who were better off and with whom there was contact; but it was unlikely to be true of many of those who lived on the peripheries, who were poorer, and with whom there was no contact.

In the third example, a senior official in a ministry in a capital city stated that in his rural home area no one ever went short of food. But a social anthropologist working in the area reported families seriously short of food during the annual hungry season; twice women were interviewed who said they had not eaten for three days. There was, however, food in the shops nearby, giving the impression that there was no reason for anyone to go hungry.

Perhaps this phenomenon is world-wide, as marked in rich urban as in poor rural agricultural society. Compared with others, the poor are unseen and unknown. Their deprivation is often worse than is recognised by those who are not poor.

Finally, we may note additional factors often missed by rural development tourists, local-level staff and even researchers. It is not just a case of the invisible poorer people. There are also other invisible dimensions: international influences on rural deprivation; social relations (patron-client, indebtness, webs of obligation and exploitation); and trends over time. The very act of being in a rural area and trying to learn about it creates biases of insight and interpretation towards what can be seen; and the observer's specialisation increases the likelihood of one-sided diagnoses, explanations and prescriptions. Poor people on disaster courses may not be recognised. A nutritionist may see malnutrition but not the seasonal indebtedness, the high cost of medical treatment, the distress sales of land, and the local power structure which generate it. A doctor may see infant mortality but not the declining real wages which drive mothers to desperation, still less the causes of those declining real wages. Visibility and specialisation combine to show simple surface symptoms rather than deeper combinations of causes. The poor are little seen, and even less is the nature of their poverty understood.

Notes

1 The male-biased syntax is deliberate and descriptive. Most rural development tourists are men.

2 Another problem is the cavalcade. The more the layers of hierarchy – international, national, regional, district, subdistrict – and the more the departments and institutions involved, so the number of vehicles increases. This adds to dust and mud if the tarmac is left, and to delay even if it is not. The record is held by a visit in Indonesia to inspect a road being financed by USAID. Douglas Tinsley reports that there were 47 vehicles involved. Ferries had to be used where bridges were not complete. At one ferry it took three hours to get the whole procession across. But there was a positive side, one supposes. The christening of the road was substantial, and the visitors cannot have been too rushed in their inspection of the quality of the roadwork, at least near the ferries.

3 'They do not know that there are living people here.'

4 This does not necessarily reflect adversely on the choice of villages, since peri-urban villages, like any others, are a legitimate subject of study.

5 An early example is provided by Zambia's fuel shortage which led to fuel rationing, following Rhodesia's unilateral declaration of independence in 1965. One effect was that the Universities of Nottingham and Zambia joint research project concerned with the productivity of agricultural labour was restricted to work in two areas instead of three, and these were areas which were relatively well-developed agriculturally, having had large inputs of education, extension and communication (Elliott, 1970, p. 648).

6 Or close to the famous tourist site for the VIP, such as the Taj Mahal at Agra in India. J. K. Galbraith has written that as hopes and enthusiasm for rural community development in India waned, 'A number of show villages continued to impress the more susceptible foreign visitors'. He records this incident:

> In the spring of 1961, Lyndon Johnson, then vice president, was taken to see one of these villages in the neighbourhood of Agra. It was, of the several hundred thousand villages of India, the same one that Dwight D. Eisenhower had been shown a year or two before. It was impressive in its cleanliness, simple cultural life, handicrafts, and evidence of progressive agricultural techniques. Johnson, an old hand in problems of agricultural uplift and difficult to deceive, then demanded to see the adjacent village a mile or two away. After strong protesting words about its lack of preparation to receive him, he was taken there. This village, one judged, had undergone no major technical, cultural, or hygienic change in the previous thousand years.
>
> (1979, pp. 106–7)

7 *Mea culpa*. In the 1960s so many of us students and other researchers were attracted to work on the (well-documented, well-organised and

well-known) Mwea Irrigation Settlement in Kenya that farmers complained about interview saturation.

8   In February 1979, two British Members of Parliament visited the Anand Cooperatives in India. They saw and were impressed by the delivery of milk from small producers to one centre. Inside hung a photograph of James Callaghan, the British Prime Minister, taken during his visit to the same centre. Asked if they would like to see a second centre they readily assented. Once inside they found another photograph, this time of the visit to that centre of Judith Hart, the British Minister of Overseas Development.

9   For the statements in this paragraph see Longhurst and Payne, 1979; Chambers, 1979; and Chambers, Longhurst and Pacey, 1981.

10 See Chapter 5.

CHAPTER TWO

Two cultures of outsiders

'Do you know what he's talking about?'
'I haven't the least idea.'
            Conversation recounted by C. P. Snow in *The Two
            Cultures and the Scientific Revolution*

Outsiders polarise into two cultures: a negative academic culture, mainly of social scientists, engaged in unhurried analysis and criticism; and a more positive culture of practitioners, engaged in time-bounded action. Each culture takes a poor view of the other and the gap between them is often wide. The two cultures partly coincide with two clusters of explanation of rural poverty: 'political economists', mainly social scientists and academics, explain it primarily in terms of social relations; while 'physical ecologists', mainly natural scientists and practitioners, explain it primarily in terms of physical and biological factors. A balanced view may best be sought in a pluralism which straddles both academic and practitioner cultures, which accepts both social and physical explanations, and which is open to the third culture, of rural people in a particular place.

Outsiders not only observe, or fail to observe, rural poverty; some also analyse it, and some try to act on it. To assess the analysis and action, we need to look more closely at the background, conditioning, values and experience of those who analyse and those who act. Separating out and describing groups of outsiders is bound to do violence to the subtleties and overlaps of reality; nevertheless, trying to be brief and clear, I shall in this chapter describe two contrasting cultures of outsiders, and two clusters of interpretation of rural poverty. I shall argue that each culture and each cluster is incomplete, giving only a partial view, and that through pluralism – a synthesis of the two cultures and of the two clusters – analysis will come closer to the truth and actions can be identified which will be better suited to needs.

Two cultures

It was in Britain, over 20 years ago, that C. P. Snow popularised the idea of two cultures, one of scientists and one of literary intellectuals. Though himself at home in both, he constantly felt that he was moving among two groups – 'comparable in intelligence, identical in race, not grossly different in social origin, earning about the same incomes, who had almost ceased to communicate at all' (Snow, 1959, p. 2). They had a curious distorted image of each other; and there was between them a gulf of mutual incomprehension, even hostility and dislike.

Similar gulfs are found among those concerned with rural development in the Third World, but who are outsiders, being themselves neither rural nor poor. Gulfs of incomprehension, even hostility and dislike, exist between disciplines, professions and departments, and between headquarters and the field. They are also experienced between foreigners, with their distinct life styles, access and resources, and nationals with theirs. But these obvious and well-known problems distract attention from a less clear-cut but more general and enduring cleavage. This is the divide among rural development outsiders between those who analyse and those who act, between academics and practitioners. At one pole we have academic social scientists preoccupied with the 'What?' and 'why?' of development and underdevelopment, with political economy, especially who gets what, why and how, and with the processes which they see as determining the answers; and at the other pole, we have practical administrators and technical scientists who concern themselves with the 'how?' of development, with trying to change things, and with trying to get things done. The physical, linguistic and experiential distance between these two groups, each with its own culture and mores, is wide; and often there is little sympathy or communication between them. To hear a seminar in a university about modes of production in the morning, and then attend a meeting in a government office about agricultural extension in the afternoon, leaves a schizoid feeling; one might not know that both referred to the same small farmers, and might doubt whether either discussion had anything to contribute to the other.

The depth of the division is reflected in the way extremists in each culture view the other. To some critical and intolerant academics, practitioners are narrow-minded philistines and at best naive reformists, part of a system of exploitation of which they are largely unaware, while technical scientists, for their part, serve their own class, producing technologies which are not for the poor. Administrators and scientists rarely ask the key question – who gains, and who loses? – and if they do, their answers are unlikely to make much difference to the poor. In any case, in the view of some critical academics, historical forces determine patterns of change, and one can do little more than watch and analyse as inevitable processes work themselves out. The rich, powerful and urban classes will prevail, at least for a time, and administrators and scientists are their willing, if often unconscious, tools.

To some narrowly practical administrators and scientists, academic social scientists appear to indulge in esoteric and incestuous debate, muttering to one another in private languages. These academics do not understand the constraints of the real world. They criticise but do nothing constructive. Given their incompetence, this is as well; and whenever they do get involved in programmes and projects, they only make trouble. They are incapable of writing anything short or clear, or of meeting deadlines. They question priorities instead of getting on with the job. They look for things going wrong; they write about failures not successes. It is best that they stay quarantined in their ivory towers and do nothing worse than mislead the young. Rural development is hard enough without them around to make it harder.

More moderate and sympathetic views can regard the actors in each culture as conditioned by their training, environment and work, and can see how these pull them apart and make it difficult for them to meet and communicate. In this spirit, let us examine each culture in turn.

Negative academics

Academics are trained to criticise and are rewarded for it. Social scientists in particular are taught to argue and to find fault. University staff spend much of their time assessing essays, seminar presentations, and examination papers. Their mental set is evaluative. When it comes to rural development, they look for faults. Their peers, too, award them higher marks for a study which points to the bad effects of a project than one which highlights benefits. Some social scientists have, in any case, an anti-government ideology and set out from the start to muck-rake. A supposedly successful project is a red rag to some academics, a challenge to see whether it can be turned into a failure by finding hidden harmful efforts or errors which officials try to conceal.

These critical attitudes have made an enormous contribution to the understanding of rural development. Looking back over the past three decades, we can see that without such attitudes it would have taken much longer for the development professions to recognise the deficiencies of the 'trickle-down' approach, the tendencies for local elites to capture the benefits of 'development', the misery of many poorer people, and the plight of so many women. These are just some of the general insights which have led to better planning, design and implementation in rural development.

Negative interpretations can, however, mislead. They may be the result of selective perception, recording and writing, of choice of research topic, and of delays in analysis and publication. Social scientists often prefer, and can most easily get funding for, research on programmes or projects which are new. Because they are new, they are precisely those where most is going wrong; and the studies occur before the main early lessons have been learnt and corrections made. In East Africa in the 1960s research concentrated on settlement schemes and cooperatives, both new and exciting programmes of rural development to which governments and donor agencies alike attached priority. In India in the late 1960s and early 1970s research concentrated on those few districts where the Intensive Agricultural District Programme was being implemented and the green revolution was either occurring or thought likely to occur (Harriss, 1977, p. 30 ff). The outcomes are well known. Many settlement schemes in Africa were found to be expensive, to create privileged settler groups with dependent attitudes, to disappoint in agricultural production, to be staff-intensive, and to represent a misallocation of resources. For their part, cooperatives were found to be inefficient, to be captured by local elites to the exclusion or exploitation of smaller farmers, poorer people and women, to suffer poor repayment records, to fail to pay their members, and to have a high turnover of both staff and funds which tended to disappear simultaneously. The green revolution in India was less negatively reported, but it was found that increases in production mostly benefited the rich, the landlords, the merchants, and the owners of tractors and tubewells, while the poor and the landless gained less, or did not gain, or lost (UNRISD, 1974; ILO, 1977).

These studies were salutary, but often late. A few were available soon after the research. Usually, though, there were long lags between fieldwork and publication. The popular culprits were foreign researchers working for their Ph.D.s who mined their data in the peripheral Third World and then removed it to a metropolitan core for processing and adding value for themselves, returning the final product to the Third World country only, if at all, much later. But long time lags are not limited to researchers who are foreign. The problems are more general. Delays in data-processing, analysis, and writing reports, theses and books, mean that research findings come on to the market three, five or even ten years after the original fieldwork. The larger the surveys and the more the disciplines involved, so the longer are the delays. Thus while careful critiques were being prepared of the early problems of settlement schemes, cooperatives, and the green revolution, some of those problems were being tackled. By the time the criticisms were published, they did not always apply with the same force.

One example can illustrate the point. Studies of settlement projects in East Africa in the late 1960s found many faults. This was a time when many of these projects had expatriate managers, when there was a tendency to provide too many services and subsidies for settlers, and when some settlement projects could plausibly be criticised as total institutions (Moris, 1967; Gosselin, 1970; Sokiri, 1972). But many of these defects had, by the mid-1970s, been reduced or eliminated. For example, the settlements in Tanzania for Bahutu refugees from Burundi were far from total institutions, allowing the settlers much freedom of movement and autonomy in decision-making. But when a seminar for staff of the UN High Commissioner for Refugees was organised to consider the relative merits of such 'organised' settlements as these, and what was known as the 'spontaneous integration' of rural refugees (meaning that they had to fend largely or entirely for themselves), social scientists whose views had been formed in the earlier period criticised organised settlement. They did not know that many lessons had been learnt and incorporated in new settlements.

In this instance, as in others, negative social science was out of date. Often, indeed, the rate of rural change is so fast, the coverage of research so low, the traditions, methods, and questions of research so conservative, and the processing of research so slow, that social scientists are permanently behind the times, failing to keep up either with rural conditions or with government practice.

Development studies are, moreover, permeated with gloom and doom. As taught in some universities, they depress. If economics is dismal, development studies are morbid. The study of historical processes itself generates a sense that things could not have been otherwise. When this is backed by an ideological framework of historical inevitability, and combined with preoccupation with what has failed or is bad, pessimism and impotence follow. This leads some political economists to undervalue even their own work (not, be it said, a notorious characteristic of social scientists generally). Thus Colin Leys, concluding his book *Underdevelopment in Kenya*, wrote that academic studies

> ...can contribute little to the effort to achieve new strategies of development grounded in the interests of the mass of those who are currently the victims of underdevelopment. Perhaps the most such studies can do is to try not to obscure the structures of exploitation and oppression which underdevelopment produces, and which in turn sustain it.
>
> (1975, p. 275)

Such pessimism understates, for a start, the impact of his own book. Critical analysis is good when it reveals processes in ways which change thinking and practice to benefit the poor. The danger is that it becomes inbred, loses touch with reality and practice, and then degenerates into sectarian narcissism and scholastic squabbles as social scientists speak less and less to anyone but themselves. The development administration movement of the 1960s and the mode of production debates of the 1970s both invite questions: who gained and who lost? The academic analysts? The poor? And did the mass of those who are currently the victims of underdevelopment deserve, and could they have had, something better?

Positive practitioners

Many of those actively engaged in planning, programmes, projects and R and D for rural development are more positive. They are those who work in government departments in headquarters and the field, technical researchers, the staff of voluntary agencies, and personnel of donor organisations. They are responsible for decisions, for acting or not acting, for allocating resources, for choosing what to do or not to do. The archetypal academic is cocooned, isolated from the operational world, disciplined only by the teaching timetable. The typical practitioner, in contrast, is more exposed, and tied to the deadlines of budgets and seasons, to targets, and to political demands. Practitioners have a sense, too, that their actions or non-actions make a difference. So while academics seek problems and criticise, practitioners seek opportunities and act. Academics look for what has gone wrong, practitioners for what might go right. Thus, for example, to contrast with Leys' conclusion, here are the final words of the summary of a book, the principal author of which was the President of the International Agricultural Development Service:

> And, while the food-poverty-population problem is massive and complex and will be extremely difficult and time-consuming to resolve, the existence of new capabilities provides a magnificent opportunity, perhaps a fleeting one, to deal with it effectively – if governments have the wisdom and the will to act.
>
> (Wortman and Cummings, 1978, p. 14)

The sobering qualification of 'if governments have the wisdom and the will to act' is almost an afterthought; the style and thrust are positive and optimistic. It is a far cry from trying not to obscure something bad to seeing a magnificent opportunity to do something good.

The dangers of positive optimism are, however, as great as those of critical pessimism. Four deserve mention.

First, energy and success over-reach themselves. In rural development practice, those who are rewarded and promoted include some who are energetic, enthusiastic and entrepreneurial. These are the sort of people who make things happen – swaying meetings, raising funds, and inspiring loyalty. These very virtues carry with them dangers of deception. Drive and enthusiasm passed down a hierarchy can generate an upwards flow of information which misleads. Vigorous programmes have to achieve their targets, so their targets are reported achieved. So it was with the areas said to be growing High-Yielding Varieties of paddy during the green revolution drives of South Asia: in a district in Tamil Nadu, the area reported was over three times the actual (Chinnappa, 1977, p. 96), and in part of Bangladesh Hugh Brammer reports that it was five times. The manner in which successful projects take off into self-sustaining myth has already been noted (pp. 17–18). And the outstanding and charismatic leader creates his or her own problems of replication. What worked quite well at Comilla under the leadership of Akhtar Ahmed Khan, could not be extended successfully to the rest of Bangladesh, nor indeed sustained fully after his departure. Replicable models are the exception, not the rule.

Second, positive practice is often unable to accept or use discordant information. Many rural development projects and programmes are delicate and unstable, especially in their early stages, and vulnerable to attack by political enemies who search for ammunition. Evaluation is called for but may be regarded by managers as threat not support, while evaluators appear as spies not allies. When criticism is offered, or damaging information comes to light, there are several possible reactions. The hardest is to accept it and change course; others are to deny it, to try to keep it quiet, to buy off the critics or to coopt them into a public relations role. Morale may then be maintained, and selective perception and myth have their part to play in maintaining support, élan and momentum; but in the long term, the costs in benefits foregone and in eventual disillusion may be high.

Third, the objectives of practitioners are often narrow. There is a recurrent tendency to home in on a single, preferably technical and physical, objective. The most common is to increase food production (regardless of whether poor people can grow the food for themselves or buy it). Others include the physical targets of construction projects, or the output targets of training programmes. Narrow professionalism here combines with practical imperatives, the need to do something and to be seen to have done it, regardless of who gains.

Finally, over-optimistic estimates are made in the early stages of projects and programmes. Partly this is because over-estimates are needed to get proposals accepted in the first place; partly because of the ease with which social cost–benefit analysis can be manipulated to produce whatever internal rate of return is thought necessary to get agreement and funding.[1] This is not an entirely negative point, however. In Albert Hirschman's theory of the Hiding Hand, habitual underestimates of difficulties which will be encountered in implementation are offset by equally habitual underestimates of the creativity which can be mustered to overcome them (Hirschman, 1967). Vision and hope are needed for action. Rural development is so difficult that some self-delusion may help to get things going at all.

Rural poverty explained?

The two cultures have contrasting ideas about the causes of rural poverty. These contrasts are far from absolute, and there are many overlaps and exceptions. Both might agree that past poverty tends to perpetuate itself. But going beyond this, the negative social science pole attracts and sustains those who explain poverty in social, economic and political terms, while the positive practitioner pole attracts and sustains those who explain it in physical and ecological terms. These views tend to cluster rather than to be absolutely separated. So we may talk of a 'political economy cluster' of views, found mostly, but not only, in the academic culture, and of a 'physical ecology cluster' of views, found mostly, but not only, in the practitioner culture. The political economy cluster sees poverty primarily as a social phenomenon; the physical ecology cluster sees it primarily as a physical phenomenon.

'Political economists', so defined, include most academic social scientists. They are so divided by discipline and by ideology that to lump them together appears, and is, simplistic. But the sharpness of their disagreements hides a premise which most of them share: that poverty is to be understood primarily in terms of economic forces, social relations, property rights, and power. Their stance can be illustrated by C. T. Kurien. In his book *Poverty, Planning and Social Transformation* he views poverty as deprivation, but not only deprivation. It is deprivation for the many and affluence for the few. He regards poverty as

> the socio-economic phenomenon *whereby the resources available to a society are used to satisfy the wants of the few while the many do not have even their basic needs met*. This conceptualization features the point of view that poverty is essentially a social phenomenon and only secondarily a material or physical phenomenon.
> (1978, p. 8. His emphasis)

Physical ecologists, for their part, are mostly practitioners and scientists. Their position can be illustrated from two authors: J. S. Kanwar and Norman Myers. In his Presidential Address to the 12th International Congress of Soil Science, Kanwar said that 'mankind today is faced with many challenges, the biggest being food shortages and environmental degradation, both resulting from the population explosion and poor resource management'. The key to the world food problem and environmental problem was better soil resource management, which was of great urgency for the survival of mankind (Kanwar, 1982). Throughout the address, Kanwar stressed physical factors such as soils, environment and population. Myers, in a brief World Environment Report, *Analysis: why the fight against hunger is failing all across Africa* (n.d.) attributes famine to natural disasters, civil disorder, adverse weather, refugees, bureaucratic problems in food distribution, man's overloading of semi-arid environments, and failure to produce more food. The emphasis is on the physical dimensions of natural and man-made disasters. Neither Kanwar nor Myers considers power, property, social relations or income distribution.

These summaries no doubt do less than justice to the three authors. Kurien goes on to gloss his point; Kanwar was a soils scientist speaking to soils scientists; and Myers might well argue that in a very short article a comprehensive statement of cause is neither possible nor needed. The fact remains, though, that Kurien on the one hand, and Kanwar and Myers on the other, presented very different views of much the same basic human problems.

Let us now elaborate this contrast more generally, and try to summarise these two clusters of explanation – of political economists like Kurien, and of physical ecologists like Kanwar and Myers.

Political economists

In the political economy cluster, rural poverty is seen as a consequence of processes which concentrate wealth and power. Although within this cluster there are many schools of thought and assertion, their differences are exaggerated by sectarian concepts and jargon and by polemical style so that it is easy to overlook the extent to which they overlap. In general, they agree that the processes which concentrate wealth and power operate at three levels: internationally, the richer countries have made and keep the poorer countries relatively poor through colonial exploitation and post-colonial unequal exchange, and at the same time benefit from the investment of capital and the expatriation of profits; internally, within the poorer countries, urban and especially urban middle class interests gain at the cost of rural interests, through shifts in the rural-urban terms of trade (cheap food for the towns, dear goods for the countryside), and through investment in urban industries and services; and within the rural areas themselves, the local elites – landowners, merchants, moneylenders, and bureaucrats – consolidate their power and wealth. For their part, the rural poor stand to lose relatively and often absolutely through all these processes. Low prices internationally, low prices internally for rural produce, and the ability of the local elite to concentrate wealth in a few hands, especially by buying land and appropriating common resources, combine with low wages to keep the poor poor or to make them poorer.

To political economists, technology and commercialisation play a part in these processes. Capital-intensive technology (combine harvesters, tractors, modern mills, large irrigation pumps, and so on) subsidised through aid, through an overvalued exchange rate, and through direct government support – is available to those who already command credit and land. Commercialisation brings with it urban products (bread, shoes, plastic goods, pots, furniture, iron frames for building, ornaments and so on) which displace rural products. Capital-intensive technology destroys the livelihoods of labourers, and commercialisation those of artisans. Both concentrate wealth in the hands of those already less poor, and enable them to become wealthier, to buy more land, to appropriate more of a larger surplus, and to expand their trade. These processes also weaken traditions of mutual responsibility and sharing, both vertically between patrons and clients, and horizontally between small farmers and between the landless. Social relationships with obligations give way to cash relationships without obligations. The rural-urban links of the wealthy are strengthened and their position is reinforced through alliances with political leaders and with the bureaucracy; poor families are weakened, impoverished and isolated.

In this view, then, the rich and powerful get richer and more powerful; and the poor become relatively and often absolutely poorer and weaker. Exceptions are recognised, such as Korea where rapid growth has benefited almost all, though some much more than others; and Taiwan where a land reform gave unusual equality in landholdings and where growth has also been remarkable. But these are seen as aberrations from the general tendency for change to concentrate power and wealth in the hands of the few at the cost of the many. Attention is directed to South America, South Asia, Indonesia, much of Africa, and elsewhere, where many of the rural poor are believed to have gained little through the changes of the development decades, or to have lost out through these processes.

Physical ecologists

In the physical ecology cluster, rural poverty is interpreted more in terms of what is physical, visible, technical, and statistical commonsense. The two most commonly cited causes of poverty are population growth and pressures on resources and the environment. Populations in developing countries are rising at 2.2 per cent per annum (World Bank, 1981, p. 108) without any prospect of an early sharp decline. In Sub-Saharan Africa the rate is higher, 3.0 per cent, and in Kenya 4.0 per cent (at which rate it doubles in 18 years). While some of this increase is absorbed in migration to towns, it is argued that urban employment and the urban informal sector are becoming saturated. Much of the increase in population will have to be supported in the rural areas. But there, under the pressure of population, land is becoming scarcer. Small farms are subdivided on inheritance and children are then poorer than their parents were. Labour supply exceeds demand and real wages go down. Some migrate to the towns to swell a miserable urban proletariat. Others move to marginal environments – steep slopes, low rainfall savannahs, and areas prone to flood or drought – where they contrive temporary and precarious livelihoods. Others compete for common resources which they decimate or destroy – fish, grazing, groundwater or forests. Fallows shorten and fertility falls. The creeping desert, soil erosion, floods, siltation, declining primary production, dropping water tables, and lower crop and livestock yields all follow. Uncontrolled population growth and uncontrolled exploitation of natural resources, in this view, combine in a vicious circle: the more people there are, the more they destroy the long-term potential of fragile environments, and the poorer this makes them and their descendants.

Physical ecologists also see the physical characteristics of poor people as explanations of their condition. Parasites, diseases, malnutrition, insanitary conditions, poor housing, lack of amenities – these are proximate causes. Physical weakness interacts with other disadvantages to perpetuate poverty. Acute shortage of food impairs the mental development of the child. Underfed people have stunted bodies. Poor people are locked into a syndrome of physical deprivation.

Poverty is also explained by climate. The correlation between mean monthly temperature and poverty is truly astonishing (Harrison, 1979 a and b[2]). Almost all the poor countries lie between the northern and southern isotherms of 20°C, and almost all the industrialised countries outside it. In one view (Harrison, 1979b) a critical factor is that up to 20°C humus forms faster than it is broken down, enriching the soil with nutrients and improving its structure; but above 20°C the bacteria work faster than the supply of dead vegetation, making it hard to sustain fertility. Other factors are also postulated in the climatic explanation – heat discouraging physical work, intensity of rainfall, prevalence of pests and diseases, and the seasonal interaction in the tropics of concurrent adverse factors – with food shortage, need to work, and diseases all coming at the same time during the rains (Chambers, Longhurst and Pacey, 1981). Moreover, some natural disasters – floods, droughts, cyclones – and the famines which follow them, are common in the tropics.

Finally, to complete the listing, there is a cluster of factors which both schools can hardly fail to recognise: war, civil disturbance, and persecution. Rural refugees who have crossed international boundaries are among the poorest and most powerless people in the world. Displaced persons within countries with civil disturbances may be even worse off because of their lack of physical protection. The tragedies, over the past ten years, of millions of rural people displaced by regimes, wars and disturbances in Angola, Zaire, Mozambique, Namibia, Rhodesia (as it was), Guinea-Bissau, Burundi, Ethiopia, Somalia, Uganda, Equatorial Guinea, the Spanish Sahara, Laos, Vietnam, Kampuchea, Burma, Bangladesh, Afghanistan, and elsewhere, speak for themselves.

Partiality

Now it may seem beyond debate that there is truth in all these explanations. But if so, why are so many of them so often left out? In practice, in both academic and practical cultures, among both political economists and physical ecologists, most analyses and prescriptions are partial, concentrating on one or a few explanations and actions and ignoring others. Three reasons can be suggested for this partiality.

The first lies in the ideological rejection by present day research of colonial beliefs. Colonial explanations of rural tropical poverty included race, climate and contingencies. The racist belief was that the natives were inferior, stupid, lazy, improvident and dissolute. The revulsion against this racist belief has been so strong and so universal that climate and contingencies have been thrown out too, tainted by guilt by association. As Gunnar Myrdal has observed:

> In the apologetic colonial theory, climate was given an important role as an inhibitor of development but this view has now almost disappeared from the economic literature. I mention this as an extreme example which demonstrates how unsparing was the post-war revolt against the colonial theory. . . . Even the word 'climate' has almost disappeared from the literature, except in occasional brief statements that climate is of no importance to development.
> (1979, p. 27)

Similarly, some of the contingencies which make people poorer, such as dowry, sickness, and drunkenness were major preoccupations of earlier analysts like Malcolm Darling (1925). The reader of the 1980s will find some of his interpretations offensive, but this does not mean that all his observations are invalid.[3] It is rather that the colonial and racist associations of climate and contingency explanations have been responsible for their premature and scientifically unwarranted burial.

The same point can be made with soil erosion in Africa. The colonial scare about soil erosion fitted with the view of shifting cultivation as backward, feckless and irresponsible. Some colonial agricultural officers sought to impose order, discipline, straight lines, and control on what they saw as a primitive and chaotic system of cultivation which was destroying its environment. The measures of control which this order was believed to require may have met the psychological needs of the agricultural officers more than the physical needs of the cultivators. In Tanzania and Kenya, authoritarian soil conserva-tion measures were so unpopular that they helped to generate and unite political organisations which then struggled for and won independence. In the post-independence decade it was scarcely then a matter of surprise that soil erosion was a non-subject. It was simply not seen or, if seen, not mentioned; thus demonstrating the astonishing human capacity diplomatically to repress and ignore discordant facts.

A second reason why some explanations are left out is the inbreeding of disciplines and professions. This point is too obvious to labour. Most university education is a narrowing, a focusing of attention and a refining of skills which exclude in order to specialise. Disciplinary academics and practising professionals meet, listen to and argue with those of similar backgrounds. A soils scientist finds his fellows among other soils scientists, or physical or perhaps biological scientists, but scarcely among sociologists; a political scientist meets and discusses with other political scientists, or other social scientists, but scarcely with research agronomists. It is not strange that there should be little overlap in their views of the problems of rural development. All have been conditioned to focus on a few aspects to the implicit exclusion of others; and members of each specialised group reinforce each others' narrow vision.

A third reason is the desire and need to simplify. Positive practitioners can only tackle one or a few factors at a time, and tend then to disregard others. Some academic analysts oversimplify, searching for or grasping at a unifying theory with simple components. More generally, syntax and the limitations of thought and speech force simplification. Writing and speech are linear and sentences (mercifully) finite. Long lists bore. There is a temptation, for the sake of coherence, intelligibility and interest, to attribute causal primacy to one factor or another – colonialism, capitalism, culture, class, greed, technology, population, low productivity, climate, environment. The pressure to shorten and simplify is especially acute with the media. As Susan George has complained: 'Swelling numbers in the poor world are ... the media's favourite – in fact sometimes only – way of explaining to readers and viewers the reasons for world hunger' (1976, p. 53). The producer of a half-hour television programme may be constrained to find one clear message that even slow-witted viewers can grasp and find of interest.

If several causes are admitted, discussion becomes more difficult, with questions of weights and primacy. An illustration is the debate over the significance of population growth. The extreme neo-Malthusianism of Paul Ehrlich (1968) is too dated to be taken seriously. But we can contrast the more measured views of Erik Eckholm, of Keith Griffin and Azizur Rahman Khan, and of Rodolfo Stavenhagen. Considering El Salvador, Eckholm acknowledges that unequal land ownership was a major contributor to environmental stress, but states that 'The obvious catalyst of El Salvador's environmental decay in this century has been the dizzying growth in human numbers' (1976, pp. 168, 183). Griffin and Khan writing on poverty and landlessness in rural Asia have a different emphasis:

> Given the structure of the economy ... and the resulting nature of the processes of capital formation and innovation, the faster is the pace of expansion of the population and labour force, the stronger will be the tendency for the standard of living of some groups or classes to fall. ... It is important to underline, however, that the cause of increasing poverty in Asia is not an alleged population explosion. Rapid population growth is merely a contributing factor. The basic causes are the unequal ownership of land and other productive assets, allocative mechanisms which discriminate in favour of the owners of wealth, and a pattern of capital accumulation and technical innovation which is biased against labour.
>
> (1978, p. 301)

Stavenhagen writes this:

> It is ... a mistake to attribute the depletion and misuse of local resources, as some authors do, exclusively to the demographic pressure on the land. While population growth has undoubtedly played a role in this process, the development of market relationships is surely the main cause of the increasing disequilibrium between population and resources at the local level.
>
> (1977, p. 45)

The words used by these writers – 'obvious catalyst', 'the cause', 'contributory factor', 'the basic cause', 'surely the main cause' – reflect their judgements and attempts to make balanced allocations of weights, in each case avoiding a narrow, unicausal dogmatism. But the words they (and indeed all of us) use imply nuances and raise questions. Is a catalyst a cause? What is the difference between a contributing factor and a basic cause? What makes something a main cause? There is also a further vocabulary – of causes which are necessary or sufficient, of conditions which predispose, of thresholds which are passed, of events which precipitate – which can be used to refine and sharpen discussion.

Partiality matters because narrow analysis and action are liable to be misdirected. Moreover the analysis and the action have forwards and backwards linkages, each reinforcing the other. Political economists who see social causes of poverty may seek to change social and property relations; physical ecologists who see physical causes of poverty may seek programmes of soil conservation, family planning, and resettlement. But the realities which these confront have complex differences – spatial, ecological, social, economic, political and temporal. The best interventions will vary. The truth is that there are many causes of rural poverty; that it is difficult to judge to what extent one or more may be primary; that the balance of their significance varies over time, by season, and by country, region, community, village, household and individual; and that not only causes of poverty but also opportunities for wealth are points of departure for rural development.

Partiality matters all the more if it is ahistorical. Rural society appears to be but is not static. Processes of change are often subtle, elusive, and rapid. Guy Hunter expressed a sense of this when he wrote of his book *Modernizing Peasant Societies*, that

> ... perhaps its central theme is that growth takes place as a long chain of small, related sequences, each of which determines the possibilities for the next. Like flood-water spreading on uneven ground, the runnels of change divide and coalesce again, here are diverted by a hillock, there are checked in a depression, build up, and break out again in new directions. The history of man in society can give some helpful indications of its general course; its detail is a matter of exact and patiently accumulated knowledge of the social topography in each case.
>
> (1969, p. 293)

This is also a warning of diversity and of the limited scope of generalisation. Any practical theory will be more valid in some places and under some conditions than others. Approaches based on partial preconceptions – whether of political economists, physical ecologists, or any one discipline, profession or school of thought – are likely to miss significant truths and fail to find the best interventions. A balanced pluralist approach, empirically based and with a wide span in both political economy and physical ecology, is more likely to fit the reality and reveal what best to do.

Pluralism

This pluralism is an ideology based on doubt, puzzlement, and agnostic openness to evidence and argument. It seeks enlightenment in both poles of contrary views, in practice seeing error less in what people say than in their condemnation of what others say. It is multidisciplinary by commitment. It requires political economists to ask about material conditions (climate, population, the environment, human physical conditions), and physical ecologists to ask about social relations (wealth, power, patronage, who controls and benefits from what, and who gains and who loses from what); and all to ask repeatedly what should best be done.

Pluralism recognises multiple causation, multiple objectives, and multiple interventions. It is suspicious of unicausal explanation, of the single objective, and of the one solution. It sees, rather, rural development in terms of many dimensions, of complementarities or trade-offs between objectives, of sequences and mixes of interventions. The logic of cost-effectiveness requires it at the same time to seek simple actions. The struggle is to improve blurred approximations to understanding, and fumbling attempts to make things better. This requires exposure to rural reality and openness to new information. It entails allowing observation and unexpected details to qualify and generate theory, rather than for theory to limit what is noticed and considered relevant.

Pluralism in the spirit of this book also embodies a philosophy of reversals: reversing the narrowing of professional vision to broaden it, reversing the direction of teaching and learning to enable core people to learn from those who are peripheral, reversing the flows of information in organisations so that those at the top learn from those below. Reversals such as these are sought, not as absolutes, but to a degree – to offset built-in biases and to achieve a more balanced understanding and better action.

Pluralism in rural development is also a way of life. It demands straddling – between academic analysis and practical experience and between the social questions of political economy and the material questions of physical ecology. This is not easy. Academics who teach gradually distance themselves from the operational world; committed practitioners are drawn away from university life and thought. Within institutes, there is a polarisation between those who lack practical field experience and who teach how dreadful it all is, and those who become involved in consultancy and are seduced by the excitement, responsibility, and flattery of 'the real world'. Those who shift from foot to foot may lose their balance; those who try to stand on both poles risk hernia. Some whose negative analysis whispers to them that the state is not benign find themselves as consultants driven to recommendations which require the benign state as a premise. A pluralist view is that this tension is creative and leads to better analysis and better practice. The tragedy is that so many outsiders are channelled into one exlusive trap or another. It is in the interests of the rural poor that they should be helped to escape: that more practitioners should spend sabbaticals standing back and seeing their work in perspective; and that more academics should be thrown in at the deep end and made to swim.

The difficulty of straddling between the academic and practical cultures varies by discipline, profession and country. Economists and statisticians have been generally successful in moving from one sphere to the other, being at ease either in a university department or in a planning office. Geographers and agricultural economists, both with hitherto rather low academic status, have made many of the best contributions. A growing band of rural development consultants contrive to have their feet in both camps. The United States has a stronger tradition than most countries of university staff being engaged in practical affairs outside the university, and this has been reflected in the involvement of United States universities in third world rural development. More generally, the rewards are often high, intellectually and in impact, of straddling the two cultures. Thus, for example, S. P. F. Senaratne, a social anthropologist, has written that the experience of working with the Sri Lanka Government

is one which I have enjoyed greatly. It has provided me with much stimulus in a setting in which disciplines such as Anthropology are poorly represented and in an academic environment which, for a variety of reasons, cannot provide that stimulus. I would go so far as to say that the Social Sciences cannot thrive in Sri Lanka until those involved recognise the present crisis in the intellectual tradition and stimulate this tradition by linking their disciplines to practical affairs.

(1978, pp. 10–11)

The third culture

Finally, pluralism in rural development has a third leg. The two cultures – academic and practical – share the top-down, core periphery, centre–outwards biases of knowledge. Both are therefore in danger of overlooking that other approach to understanding, from the bottom up, from the periphery towards the core, from the remote towards the central. For the two cultures are cultures of urban-based outsiders. The third culture, of the rural people in a particular place, is the true centre of attention and of learning. As some officials were once told, 'The village is the centre; you are peripheral'. The micro-level is again and again out of focus; and when in focus it is seen from a distance, through the urban professional's telescope. To understand rural poverty better, and to judge better what to do, outsiders, of whatever persuasion, have to see things from the other end.

Notes

1 For one agricultural project appraised by three different teams, rates of return were estimated as, respectively, 19 per cent, 13 per cent, and minus 2 per cent.
2 Harrison is, however, a pluralist, not a physical ecologist.
3 See Chapter 5.

CHAPTER THREE
How outsiders learn

'You don't know much', said the Duchess; 'and that's a fact.'
Lewis Carroll, *Alice in Wonderland*, Chapter VI

To find out about rural poverty, many outsiders use questionnaire surveys. These provide data for the planners, statisticians and economists who most easily straddle the two cultures. They also minimise the rural contact required of the urban-based professionals who use them. But questionnaire surveys often take more time and resources than estimated, enslave researchers, and generate misleading data and unread reports. Some bad questionnaire surveys make rural people appear ignorant when they are not. Other sorts of surveys, involving careful measurements and not limited to questionnaires, have much to contribute, especially when social, medical and natural scientists combine. Social anthropologists who practise total immersion in their villages learn much in depth, but are often unable or unwilling to communicate their knowledge. Examples of more cost-effective appraisal and research suggest methods which will be eclectic, inventive, adaptable, and open to unexpected information, allowing timely analysis and reporting, and involving rural people themselves as partners in research.

Urban-based professional outsiders learn about rural conditions in many ways. These methods can be loosely grouped into appraisal, which is less formal and briefer, and research, which is more formal and takes longer. At one pole there are the appraisals of casual empiricism – the explorations of the unselfconscious practitioner and rural development tourist with their anti-poverty biases (see pp. 13–23); at the other pole is the respectable research of convention – the investigations of the traditional academics with their long-drawn-out, even pedantic, reverence for correct method. In between is a middle ground, which I shall later argue presents underdeveloped potential for cost-effective learning.

Research has many origins. It is sponsored by practitioners

for their purposes and carried out by academics for theirs. It is commissioned by governments, aid agencies, foundations and private and parastatal organisations. What research is done, how it is done, how it is written up, and its consequences or lack of consequences, vary greatly. Many problems arise, but some of the most widespread and serious concern the gaps between the three cultures: between outsider practitioners and academics; and between outsiders and the rural poor.

Between the cultures

Among outsiders there is a difference between what practitioners want and what academics can or will provide. One part of this is choice of topic and emphasis. What a practitioner thinks 'useful' an academic may not find 'interesting'. Both are right in their own ways. A practitioner has a responsibility for results; an academic, for understanding. Bridging the gap, research commissioned by practitioners can exercise a healthy discipline on academics, concentrating their minds and efforts. At the same time, it is difficult to overstate the value of concerned independent observation and analysis in the traditions of critical scholarship, pursuing questions which are out of fashion, out of favour, or out of bounds. There is a danger that universities and research institutions may become too much the handmaids of governments, doing only what they are told or commissioned to do. Nothing in this chapter should be taken as undervaluing independent and heretical research. Where the rural poor are oppressed, independent writing provides one strong lever for change.

There remains, moreover, a large area of overlap between the 'useful' and the 'interesting' where governments or aid agencies commission research from universities and research institutes. And here a major problem is a difference of time scales. Practitioners usually have short time horizons. They live and work with deadlines. Government budgets and climatic seasons set dates by which information must be available if it is to be used. Decisions on agricultural pricing, on the quantities, timing and distribution of agricultural inputs, or on the estimates for next year's budget, have cut-off times for data collection. Rural emergencies – floods, droughts, refugees, water shortages, a pest or disease outbreak, an incipient famine – demand rapid assessment. Project and programme identification, too, have their time schedules; information must be gathered under pressure especially where projects or programmes have political priority.

For their part, academic researchers have longer time horizons and are less inclined to work to deadlines. There are many reasons. The worldwide acceptance of late submission of post-graduate theses teaches academics early in their careers not to take writing deadlines seriously. The secure tenure of many university posts is relaxing. Priority goes to immediate demands of lecturing, teaching, administration and university politics. The culture of university life is detached and reflective compared with government departments or voluntary agencies. The inability of most academics to manage a personal work programme pushes research again and again to the end of the queue. Finally, respectable rural research methodology requires either extensive surveys or long residence in the field, and both of these take much time. For all these reasons tension over getting results on time is almost universal between agencies which commission research and university departments or institutes which carry it out.

A further problem is the misfit between the methods and practice of research, and insight into the conditions and needs of the rural poor. The biases of rural development tourism which direct the attention of practitioners away from the poorer regions, the poorer people, and the poorer times of year, also affect researchers in the choice of locations to study. In India, a bias towards conducting research in more progressive or specially favoured areas has been documented: for family planning research by Piers Blaikie (1972); and for social and economic research by John Harriss (1977). Censuses and research using questionnaire surveys easily miss poorer households or groups who hide or who are remote; or they are omitted by enumerators as unimportant or too much trouble, or because they are those least likely to complain if they are left out. Surveys also tend to be carried out in the dry seasons, and quite rarely during the bad times of the rains before harvest; and annual averages of incomes or calorie intake mask seasonal deprivation. An exaggerated impression of general wellbeing can also be given by the averages beloved of researchers: if there is a skew distribution of incomes or wealth, the ends of the distribution tend to be obscured. For reasons which are financial, administrative, and statistical, as well as possibly political, anti-poverty biases are embedded in many official surveys (Ward, 1979).

Convergence on questionnaires

The most common method of formal rural research is the

questionnaire survey. The purpose of outsiders may be to find out about subjects as diverse as farming practices, family planning, agricultural extension, child care, nutrition, medical knowledge, household income, literacy, or use of the media, but whatever the purpose the reflex is the same. A questionnaire is drawn up, a sample selected, and the questionnaire applied.

Several forces combine to promote these questionnaire surveys. The strongest bridge between the two cultures, of practitioners and academic social scientists, has been the rubric and reality of 'planning'. To academics, planning is an acceptable activity being concerned less with instant nuts and bolts and more with policy in the medium or longer term. But planning, whether national or local, requires 'data' about rural people which can be aggregated to give an overall view. What other mechanisms for obtaining such data than surveys? Again, agencies concerned about the effectiveness of projects they have funded want to know what they have or have not achieved. What better than a benchmark survey of the project area and of a control, with follow-up surveys later? Then, many non-social scientists, and especially natural scientists, have a mathematical training, a reverence for hypothesis-formulation and testing, and a belief that the social sciences should strive for a rigour similar to that of the natural sciences. Questionnaire surveys subject to statistical analysis seem to meet these requirements. Another factor is professional predispositions in economics and statistics. Economists are better able than those in most other disciplines to straddle between practitioners and academics. They therefore unduly influence the nature and style of collaboration. Statisticians, for their part, whether in ministries or research institutes, must justify their existence; and to do this they need numbers. So economists and statisticians, both numerate, both acceptable to both cultures, and both required in 'planning', demand surveys and the statistical data which they generate, and which allow them both to consummate their professional skills and to be, or at least appear to be, useful.

Convenience, class, prestige and power also play their part in promoting surveys. The analysis of survey data can be done safely and comfortably in an urban office without rural exposure. It reinforces what M. N. Srinivas has described as 'The division of labour between the theoretician-analyst and the fact-gatherer', the latter constituting a 'helot class' which does the rural work of investigation and enumeration, allowing the analyst to work away without the inconvenience of contact with the reality (1975, pp. 1389, 1390). The manipulation of figures is a clean, tidy and unpolluting activity. Arcane mathematical mystery allows its high priests to criticise, veto or amend the research of others; and it demands sophisticated computers for its devotees.

These forces help to explain why an urban-based industry of rural social surveys has mushroomed, financed by national governments, research councils and foundations, and following the changing fashions of topics – the diffusion of innovations, family planning, the green revolution, agricultural practices, cooperatives, credit, rural industries, employment, self-help, baseline or benchmark surveys for projects and programmes, and the plight of women. Surveys are a respectable reflex. It is scarcely surprising that a 1974 Conference on field data collection in the social sciences concerned itself mostly with data collection through surveys and little with other techniques (Kearl, 1976). In the minds of some, rural research *is* surveys.

But questionnaire surveys have many well-known shortcomings. Unless careful appraisal precedes drawing up a questionnaire, the survey will embody the concepts and categories of outsiders rather than those of rural people, and thus impose meanings on the social reality. The misfit between the concepts of urban professionals and those of poor rural people is likely to be substantial, and the questions asked may construct artificial chunks of 'knowledge' which distort or mutilate the reality which poor people experience. Nor are questionnaire surveys on their own good ways of identifying causal relationships – a correlation alone tells us nothing definite about cause – or of exploring social relationships such as reciprocity, dependence, exploitation and so on. Their penetration is usually shallow, concentrating on what is measurable, answerable, and acceptable as a question, rather than probing less tangible and more qualitative aspects of society. For many reasons – fear, prudence, ignorance, exhaustion, hostility, hope of benefit – poor people give information which is slanted or false.

For these and many other reasons, conventional questionnaire surveys have many drawbacks if the aim is to gain insight into the lives and conditions of the poorer rural people. Other methods are required, either alone, or together with surveys. But extensive questionnaire surveys pre-empt resources, capturing staff and finance, and preventing other approaches. Let us examine this phenomenon more closely.

Survey slavery

The costs and inefficiencies of rural surveys are often high: human costs for the researchers; opportunity costs for research capacity that might have been better used; and inefficiencies in misleading 'findings'.

Thousands, perhaps tens of thousands, of researchers have surrendered their freedom to surveys; and if field workers are helots, their masters can also be slaves. For preparing, conducting, analysing and writing up a rural survey are heavily committing activities, the demands of which are habitually ignored or underestimated, and the duration of which almost always exceeds that planned.

Commitment to surveys is all too easily and willingly accepted. It is not just that statisticians, economists and others have professional preferences. Research institutions and universities need to obtain funds; once they have conducted some surveys, there are pressures and obligations to find further employment for field staff, who then go from project to project for years; and funding sponsors are prepared to pay for surveys because they feel that they will get at least something, an identifiable and justifiable product, for their money. Commitment then deepens. The more complicated, extensive and expensive the survey, so the more sophisticated will be its data processing (more marks for computers, programming, tapes and printouts than for anything as primitive as hand tabulation), the greater the prestige for the senior researchers, and the more time required. There is also a 'because it's there' element, a sense that until social scientists have conducted their surveys and struggled with their computers, they have not climbed their Everests. And like attempts to climb Everest, extensive surveys require much administrative and logistical support, cost a lot, and often fail.

The pathology of rural surveys follows common paths. Its demands are not properly estimated. At the planning stage, it is easy and tempting to expand the geographical area to be covered, the numbers in the sample, and the questions to be asked. Where a team is involved, with each member contributing ideas, the questionnaire grows. The more multi-disciplinary the team, the greater is the questionnaire's potential for growth: the more disciplines, the more questions. It is also easier to admit a new question than to argue with a colleague in another discipline (and with whom one has to work for months or years to come) that his or her question is unnecessary. Short-term peace in the team is bought at the cost of long-term liabilities. The outcome is excessive data to collect and therefore less likelihood that the data will be well collected or that they will be checked, coded, punched, processed, and analysed, and less chance of the distant consummation of the survey being written up, let alone read and acted on. And on top of this there are the administrative demands of recruiting and training enumerators, the logistics in the field, and the thousand and one technical and practical problems of implementation.

Whatever the problems, commitment to completing a survey is irreversible, often from the start; there is no going back. Under pressure of the immediate need to keep the survey running, its objectives slide out of sight; the means – the collection of information – become the end. There is neither time, energy nor resources to explore new questions or to notice the unexpected. Urban bias grows with unkind irony, as administrative and logistical demands tie senior researchers to urban areas and confine their brief field activities to administrative matters – housing and allowances for investigators, supplies of schedules, pay. The survey becomes a juggernaut pushed by and pulling its researcher slaves, and sometimes crushing them as it goes.

As data collection is completed, processing begins. Coding, punching and some simple programming present formidable problems. Consistency checks are too much to contemplate. Funds begin to run out because the costs of this stage have been underestimated. Reports are due before data are ready. There has been an overkill in data collection; there is enough information for a dozen Ph.D. theses but no one to use it. Much of the material remains unprocessed, or if processed, unanalysed, or if analysed, not written-up, or if written-up, not read, or if read, not remembered, or if remembered, not used or acted upon. Only a miniscule proportion, if any, of the findings affect policy and they are usually a few simple totals. These totals have often been identified early on through physical counting of questionnaires or coding sheets and communicated verbally, independently of the main data processing.

A report is required. It has to be written late, by dispirited and exhausted researchers who have already begun new tasks. Their families do not thank them for their absences, late nights, and short tempers. They stare at print-outs and tables. Under pressure for 'findings', they take figures as facts. They have neither time nor inclination to reflect that these are aggregates of what has emerged from fallible programming of fallible punching of fallible coding of responses which are what investigators wrote down as their interpretation of their instructions as to how they were to write down what they believed respondents said to them, which was only what respondents were prepared to say to them in reply to the investigators' rendering of their understanding of a question and the respondent's understanding of the way they asked it; always assuming that an interview took place at all and that the answers were not more congenially compiled under a tree or in a teashop or bar, without the tiresome complication of a respondent. The distortions are legion. But mercifully, however spurious their precision, 'findings' printed out by a computer have a comforting authority. The machine launders out the pollutions of the field and delivers a clean product, which looks even cleaner and more comfortingly accurate when transferred to tables and text. These 'findings' are artefacts, a partial, cloudy and distorted view of the real rural world. But in the report they are, they have to be, facts.

Writing the report, then, demands casuistry. Conclusions have indeed been arrived at, but they are based on anecdote, common sense, observations incidental to the survey, 'I-once-met-a-farmer-who-' statistics, and the opinions of local people and officials. But the report-writer feels obliged to derive them from the survey's formal statistical output. Cosmetic surgery on the body of data improves appearances; sloppy syntax slurs non-sequiturs; concluding paragraphs assert that the data showed, or proved, what careful reading would show they did not show, or prove, but which may be true nonetheless. As it is, no one will read the report in enough detail to notice this, for the writers have compulsively crammed it with almost raw data. They have felt that all, or most of, the data must be presented, lest all that awful effort should have been in vain. It must all, surely, have some value sometime to someone somewhere. And indeed, it has. It is there, undigested and unabsorbed. It is not read, because it has been written in execrable style: jerky, unmemorable and ugly. Tables, statistics and turgid prose cloy the reader's critical faculties. So either the text goes unread; or if read, not understood; or if understood, not remembered. This serves the report well, investing it with authority; for who can challenge the conclusions without being sure that they are not supported somewhere in the document or its appendices? Dull survey data badly written up present a background against which other information stands out; and what stand out and may be remembered are those simple conclusions gained outside the survey which, happily, are more likely to combine truth with usefulness.

Finally, after the report or the book, evaluation of the survey process is unthinkable. It has taken so long that the main actors are exhausted or have moved on. The staff in the funding agencies who sponsored the survey are now in other jobs, and their successors have other surveys planned or in progress. There is an (unread) report as a monument or tombstone for the project. At least this is something to show for the money. And in any case evaluation might be damaging because of what it would reveal, if, that is, it was more accurate than the survey itself. Honest self-criticism is neither easy, rewarded, nor popular. There is no Journal of Misleading Findings. To describe the muddle, shortcuts, and fudging might destroy the survey in the eyes of colleagues, peers and sponsors. Too much is at stake: the reputation of the institution, the career of the researcher, the chances of future contracts and jobs; or so it is believed. Criticism is not put in writing; or if it is, it is the first victim of the editor's pencil. It would be damaging in the report; or the publishers demand that the book be shortened, and since none of the findings in the chapters by individual authors can easily be cut, the section on methodology suffers most. And honesty loses friends and may be disloyal. To criticise one's own shortcomings is one thing; let the solitary social anthropologists with their disarming candour by all means continue to tell stories against themselves. But to criticise the shortcomings of a team is to impugn colleagues, perhaps friends. Better, it will seem, to remain silent. And so it is that myth masquerades as fact, unchallenged, to two places of decimals, and new innocents plunge unwarned[1] into the morass.

Misleading findings

This is a description of the pathology of surveys. Not all are so bad. There are 'good' surveys. But the difficulty is to know how good they really were. When probed, the results of supposedly good surveys are sometimes alarming. Cases of apparent under-reporting and over-reporting are many, as are motives of respondents for exaggeration and concealment. Where painstaking, sensitive cross-checking is carried out, major errors are often revealed, as five examples can illustrate:

1 R. S. Arole (1977, p. 22) found in Maharashtra in India that when professional project staff surveyed a village to establish the incidence of abortion, not a single woman admitted having had one; but the village health worker, who was closer to them, regularly reported abortions, legal and otherwise.
2 A careful investigation in Nepal (Campbell, Shrestha and Stone, 1979) showed that the Nepal Fertility Survey understated the knowledge of rural people in medical and family planning matters. The investigation compared responses given in the survey of a national sample with those obtained by a more careful survey of a much smaller sample, including a follow-up to cross-check negative responses. The results were striking and can be seen in Table 3.1. The reasons

Table 3.1 Rural people's knowledge of family planning

	Percent of positive responses according to Nepal Fertility Survey (national sample)	Percent positive according to the study survey (N=76)	Percent positive after cross-checking (N=76)
Heard of pills	12	63	97
Heard of loop	6	56	91
Heard of condom	5	45	95
Heard of vasectomy	16	58	95
Heard of abortion	5	64	100

Source: Campbell, Shrestha and Stone, p. 5
Percentages rounded to whole numbers

for the low responses in the Nepal Fertility Survey included the unintelligibility of the questionnaire which used a highly literate variety of the Nepali language, the sensitivity of the subject, and the social setting of the interview. The Survey's 'findings' led to the conclusion that ignorance of family planning was a major problem and that basic education was needed, but the real problem was crude methods of investigation. The ignorance of rural people was created by the ignorance of outsiders.

3 Mahmood Mamdani, in his book *The Myth of Population Control*, found a survey team being given and recording misleading information on the acceptance of family planning. He reports why people accepted contraceptive tablets but did not use them, as implied by a traditional medical doctor, who said of a visiting survey team:

> But they were so nice, you know. And they came from distant lands to be with us. Couldn't we even do this much for them? Just take a few tablets? Ah! even the gods would have been angry with us. They wanted no money for the tablets. All they wanted was that we accept the tablets. I lost nothing and probably received their prayers. And they, they must have gotten some promotion.
>
> (Mamdani, 1972, p. 23)

4   Sean Conlin lived as a social anthropologist in a village in Peru. While he was there a sociologist came and carried out a survey. According to the sociologist's results, people in the village invariably worked together on each others' individually owned plots of land. That was what they told him. But in the period of over a year during which Conlin lived in the village, he observed the practice only once. The belief in exchange relations was, he concludes, important for the people's understanding of themselves, but it was not an economic fact (Conlin, 1979).

5   John Harriss lived in one of twelve villages in North Arcot District in Tamil Nadu which were being surveyed by enumerators. Using his detailed local knowledge and checking field plots, he recalculated the acreages under crops reported to the enumerator in questionnaire interviews. He found that the 100 acres reported to the enumerator were in reality 150, and that of the 50 acres not reported, no less than 37 were being cultivated by the four largest farmers in the village. Rechecks in the other villages by the other enumerators identified underestimation of area covered for the twelve villages as a whole of 14 per cent at the listing stage of the survey and 8 per cent at the later farm survey stage (Chinnappa, 1977, p. 43). If similar understatements by larger farmers were to be found in all surveys in rural India, then the skewness of land distribution would have been sharply underestimated. The fallibility of land-holding data is nicely captured in what has been called Panse's Law, which can be expressed thus: the average size of land-holding in a village increases with the length of residence of the investigator (Panse 1958, p. 224). With questionnaire surveys, the residence is often short; in many there is no residence, but merely day visits, and then the average land-holdings stay small.

These examples of misleading data are not the end of the story. Debunkers are not always right; but who debunks the debunkers? There are always questions to be asked about method. In the survey to check the Nepal Fertility Survey, what happened when negative responses were cross-checked? If a person is asked if he or she has heard of a condom, and says no, and is then asked again, might the reply not be yes (because they heard of it when they were asked the first time, or the survey raised a flurry of interest in the subject so that everyone was talking about it, or because they became ashamed to continue to show ignorance)? And what of Mamdani's methods?[2] Was he, like those he criticised, told by villagers what they thought he wanted to be told? It is difficult to answer these questions, but they serve to stress the need to describe and be critical of one's methods, and always to retain some residual doubt.

Mention must be made of one more casualty of the pathology of extensive questionnaire surveys: the seasonal dimension. Seasonal analysis of data requires much painstaking work. It defers the gratification of finishing the Ph.D. thesis, report, or book. Often there is neither money, staff, interest nor patience for it. Yet there is abundant evidence that the experience of deprivation in the rural tropics is seasonal. It is rare indeed to find a study which adequately covers seasonality of deprivation in health, nutrition and agriculture.

Useful surveys

The thrust of the argument is not that questionnaire surveys should be abandoned, but that they are more limited, less reliable and less able to generate insight than is commonly believed. By capturing and enslaving so many researchers, especially social scientists, they also raise questions of cost-effectiveness and opportunity cost, of alternative uses of those same resources of staff and funds. But they remain a legitimate, necessary and useful tool, especially for data which are not sensitive and for which distributions and aggregates are needed. They can help establish some of the orders of magnitude needed in planning. Some, like the surveys of households in rural Zambia by Marter and Honeybone (1976), identify patterns of inequality and poverty which affect thinking and action about rural problems. But questionnaire surveys must be prepared and conducted with scrupulous care and wherever possible subjected to independent checks with other more sensitive and in-depth methods.

Some of the most useful surveys have involved several disciplines and have used other methods of measurement besides questionnaires. Many of the best have involved collaboration between social and natural scientists. Not least, such research can expose and dispose of myths which survive because such collaborative research is difficult and rare. To take one example, the 'development profession' of the 1960s and 1970s adopted and propagated a belief that in village conditions post-harvest grain losses were very high: the figure of 30 per cent was often repeated, and lodged itself in the minds of those who spoke at international conferences. We still await a full study of the origin of this figure, but Martin Greeley (1982, p. 33) has argued plausibly that it came from high losses in marketed grains, not in village-level storage. It was only with the most careful and well-managed field research, involving both social and natural scientists, that this myth was exploded. One team working in Andhra Pradesh in India found farm level storage losses of only 4.3 per cent (Boxall *et al.*, 1978), while another working in Bangladesh, found total physical food losses for rice at the farm level to be below 7 per cent (Greeley, 1982).

Similarly, to understand the interrelationships of disease, nutrition, social conditions and poverty, requires careful longitudinal survey work across a range of medical and social science disciplines. What sometimes happens is that excellent work in some disciplines is not complemented by any work at all in others. The value of field medical research of the highest quality, like that of the Medical Research Council in the Gambia, and of the Cholera Research Laboratory (now the International Institute for Diarrhoeal Disease Research) in Matlab Thana in Bangladesh, is greatly enhanced when adequate attention is paid to social factors. How rarely this is done is indicated by how few are the cases where rates of morbidity or mortality have been analysed by socio-economic class. The benefits can also be high where detailed longitudinal surveys in the social and agricultural sciences, like those conducted by ICRISAT in India, are complemented by work on health and nutrition. Short, *ad hoc*, investigations which 'piggyback' on major surveys, can also be very cost-effective.

The need here is for a limited number of surveys which are better, last longer, cover a wider range of disciplines, and are well analysed, and which will capture and interpret better the realities of rural deprivation by including health, nutrition and agriculture as well as the closer concerns of the social sciences.

Total immersion: long and lost?

One alternative is the approach of social anthropology. Most of those cited above who found major flaws in surveys lived in villages and gained their insights through participant-observation and questioning. There is much to commend this approach. Village residence may mean risks of overgeneralising from the particular village. But the depth and richness of insight often more than compensate for that by penetrating personal, historical, economic, social and political relationships and trends. But here, too, there can be serious drawbacks in the full, respectable, professional, approach.

This can be seen in the sequence of initiation, fieldwork and writing up of social anthropologists. They are by choice solitary. (The residential clustering of social anthropologists in the Kalahari desert is probably *sui generis*.) The normal pattern is total immersion in a village as part of the rite of passage for entry into the professional guild. Correctly performed, this part of the apprenticeship takes one to two years, during which time data are collected inclusively. Some try to avoid the taint of contact with government or government programmes, and discourage visits by other outsiders. Considerable amounts of data are amassed and rich insights often gained. But later the data can become oppressive. The longer the fieldwork, the more the data, and the greater the difficulty in writing up. Through great struggles a few papers are forced out at intervals; but sometimes their erudition is matched only by their practical irrelevance and their inaccessibility to policy-makers, who might not understand them even if they knew the journals and had time to read them. A contribution is then made to the archives of professional knowledge, but not to the alleviation of poverty. Moreover, many social anthropologists have been unable or unwilling to give practical advice. Asked for suggestions about what to do, they might give replies either of the 'that's not my department' type, or on the lines of: 'Give me five years and I will tell you why I need longer before I can tell you why you should proceed with the greatest caution.'

To this stereotype of the classical social anthropologist there are now many and increasing numbers of exceptions. Development anthropology which seeks to make a practical contribution has gained momentum and acceptance. But there are still chances lost. Social anthropologists have precious opportunities in their fieldwork. They have access to a world of experience normally shielded from the outsider. It is the very open-mindedness and lack of predetermined structure to their investigations, which makes this possible. Thus Srinivas, Shah and Ramaswamy in their introduction to a volume on *The Fieldworker and the Field*:

> . . . the fieldworker cannot anticipate the developments in the field which will inevitably guide the course of his investigations. Hypotheses formed without regard to these considerations may turn out to be trivial if not banal. Almost no contributor to this volume has been guided by hypotheses, and some confess that their theorizing was only post factum. What most fieldworkers do is to go to the field with a grounding in the theory of the discipline, especially in the sub-area of their interest, and with as much knowledge of the

region as can be derived from secondary material. The field then takes over, and the outcome depends on the interaction between the fieldworker and the field.
(Srinivas *et al.*, 1979, p. 8)

The result is that social anthropologists who have spent any length of time in the field usually know a great deal and have many insights useful to practitioners and interesting to academics. But often they do not know what they know, and their self-isolation hinders those exchanges which could most illuminate and help others.

Those who break out of conventional modes of expression sometimes have the greatest influence. Elenore Smith Bowen, after her fieldwork in West Africa, wrote a novel, *Return to Laughter*, which has been widely read, which influenced a whole generation of Peace Corps volunteers, and which has probably made a greater contribution to understanding and practice than any learned thesis she could have written. Adrian Adams, reacting to the attitudes and impact of outside 'developers' on the village where she lived in Senegal, wrote an 'Open Letter to a Young Researcher' (1979) which deserves to be equally widely read; for it describes the villagers' view of the people who came from outside, their experiences of being misled and of having their initiatives undermined, and their sane disillusion, leading eventually to this letter to a young researcher who came to the village to visit her, explaining why she would not see him. Circumstances differ so much that it is unwise to generalise. But whether through writing conventionally, anecdotally, and with passion or through enabling policy-makers to learn, or through descriptions of reality at village level, social anthropologists can do so much to help rural people – and especially the poorer rural people – that it is a tragic waste when their ritual immersion becomes sterile quarantine.

Cost-effectiveness

These two examples – of extensive surveys, and of total immersion – raise questions of cost-effectiveness in research. In each case, the method seems to dominate; conventions dictate what is done and then lengthen the period of research and writing. Professionals choose topics for research which require, exercise and consummate the skills in which they have been trained. But this means that methods and skills are looking for problems, the tail wagging the dog. Reversing this, the question is what problems should have priority. Obviously, objectives should come first, and methods only second, assessed for their cost-effectiveness in achieving those objectives. But objectives for research differ widely, and are often multiple. And even if improvements for the rural poor are taken as an overriding objective, there remain questions of what constitute improvements, who determines what they are, who the rural poor are, and what causal connections are anticipated between the research and those improvements.

In assessing cost-effectiveness, there are further imponderables. The cost side is straightforward only if interpreted narrowly in terms of finance: it is not difficult to add up the costs of salaries, transport, paper, offices, computer time and the like. But the true cost of research is less tangible. It is the opportunity cost of all the resources used, the benefits foregone from not using them in other ways. These opportunity costs are high where research staff are few. Committing staff to surveys preempts their time and expertise so that they cannot undertake more qualitative work.

The exact benefits of research are unknowable in advance; if they were knowable, the research might be less necessary. Benefits also work themselves out over many years and in many places, in changing the research priorities of others, in changing opinions, in the design and implementation of projects and programmes, and so on. It is tempting then to take refuge in the ultimate uncritical academic faith that because knowledge is good any addition to knowledge is worthwhile. With rural deprivation, where the questions concern the life, suffering and death of hundreds of millions of poor people, that view cannot be sustained. Instead, tough thinking is called for about priorities and choices in the deployment of resources.

This can be underlined by examining ways in which poor rural people can benefit from appraisal and research. Three methods in particular stand out: the direct operational use of data; changes in outsiders' awareness, knowledge and understanding, leading subsequently to changes in their behaviour; and the enhanced awareness and capability of the rural poor themselves.

One danger of research and appraisal is the concentration on some parts of the research process to the neglect of others. The law of prior bias operates – what comes first, gets most. The early stages, especially of data collection, are prominent. But cost-effectiveness requires impact. This in turn requires analysis of trade-offs between quantity, quality, and relevance of information, and then of its actual use and impact. Appraisals often have little impact because of the irrelevant or unread report. Academic work is even worse. Supposedly rigorous in research methodology, academics are astonishingly unrigorous in the diffusion of their findings. To impress their peers and promotion boards they publish impenetrable prose in prestigious journals, spurning as journalistic those papers and bulletins which take articles which are brief, clear, practical, and read. Enormous sums are devoted to research and little to diffusion of research results. Diffusion and impact are often left to take care of themselves.

There is a paradox here. Criteria of cost-effectiveness and a hardnosed approach to benefits for the rural poor may direct attention and accord priority to investigations with short rather than long causal chains and with certain rather than uncertain outcomes. Thus a survey to identify families to be affected by the building of a dam and to assist in their resettlement may be preferred to, say, case studies of the survival strategies of poor households. But it by no means follows that the benefits from the former will be greater than those from the latter: all we can say is that they are more direct and more identifiable. The long-term impact of, for example, a study of the processes whereby land ownership becomes concentrated and many people become landless may have no direct or early impact, but may be imprinted on the consciousness of a whole generation of students, some of whom, in due course, support and carry through a land reform.

Breakthroughs and new insights come in many ways, some direct, many indirect. Pursuing what is intellectually exciting may be efficient as a general practice, although its efficiency is difficult to predict in any one case. To maximise its benefits requires openness to information, lateral thinking, and an ability to notice and follow up the unexpected. For this, two extremes are to be avoided; the pedantry of the short-sighted, slow-moving snail amassing facts; and the dilettantism of the butterfly that flits from flower to flower. There are more snails than butterflies; and more snails should look up and become, if not airborne, at least more aware of their wider surroundings.

Changes in the awareness, knowledge and understanding of outsiders come in different ways. Surveys generate a very few statistics which are remembered and repeated and which have an impact. But they are dry. Minds are soon numbed by figures. Percentages of malnourished children, or per capita incomes, or numbers of children not in school, have meaning but are not deeply moving. There are also problems of credibility. Poverty may be under- or overstated. Overstatements have been made especially in nutrition surveys. For example, the Indian National Sample Survey 1961/2 'found' that 85 per cent of the population of Kerala consumed less than 2 200 calories per day, but their measurements omitted or underestimated some items of diet, including coconut, tapioca, and jak-fruit (UN, 1975). In India, the estimates for undernutrition have ranged from the estimate of Dandekar and Rath (1971) of 50 per cent of the urban population and 40 per cent of the rural, to P. V. Sukhatme's (1977) of 25 and 15 per cent respectively, with the debate continuing. The figures change and the layperson is bemused. The statistics which would more accurately delineate deprivation – morbidity and mortality by region, by socio-economic class, and by season, for example – are often not known. And in any case, surveys are poor tools for insight into relationships.

If the objective is improved conditions for the poor, then the outsider, with help from the rural poor themselves, must try to identify and understand processes, linkages, and opportunities for change. This can usually be done better through an anthropological approach open to a wide range of information, and flexible enough to follow up leads, than through the application of a predetermined survey instrument. Case studies stimulate and inform more than statistics. Unfortunately, many urban people believe that they know all about rural poverty already. The observation that ' . . . both officials and politicians seem to think that they know everything that needs to be known about rural India except statistics' (Srinivas, Shah and Ramaswamy, 1979, p. ix) applies in other countries too. There is a paucity of case studies of individual families and their strategies.[3] Outside social anthropology, these have not been regarded as really respectable outputs from research. Journalists may write up that sort of thing, it has been felt, but hardly serious social scientists. But without the rich realism of cases, it is easy for glib elitist stereotypes of the stupid, ignorant and lazy poor to persist. Where case studies of poor rural households are found, they often reveal a resilience, stamina and ingenuity which members of urban elites would be proud to recognise in their own families. Case histories of families and of individuals are one of the better ways for changing what outsiders know and feel about the rural poor.

Four ways in and out

The benefits of improvisation and inventiveness in methods of appraisal and research can be illustrated by four examples. They are not presented as ideals, but to show that very different approaches can be effective, and to see what they have in common.

i) *Ladejinsky's tourism and the green revolution*

Wolf Ladejinsky has been described as

> a major voice calling out for economic, social and technological measures to aid impoverished peasants in the developing areas. His work changed the lives of millions of people from Japan to India. No one was as important to the success of postwar land reform in Japan and Taiwan and no one worked harder to promote similar efforts in Vietnam and India.
>
> (Weisblat, 1976)

A man of wide experience, he carried out two brief field trips in India in 1969, at the age of 70, and wrote them up in the *Economic and Political Weekly* (Ladejinsky, 1969a and b). He visited the Punjab and the Kosi area in Bihar. His methods were mixed. He made use of surveys and official statistics. He talked and listened to farmers and labourers. Like all rural development tourists, he was vulnerable to distorted information. He observed of a conversation with a landlord in Bihar

> He first informed us that he owned 16 acres of land but corrected himself under the good humoured prodding of a crowd of farmers that he had failed to mention another 484 acres. The lapse of memory might have had something to do with the ceiling on land-holdings and its maximum permissible limits of 60 acres, but, on the other hand, no owner bows his head in shame on account of ceiling evasion.
>
> (1969b, p. 9)

One senses that Ladejinsky had the experience and skill to see through to the realities in spite of the limitations of his rapid and informal methods. In 1969 he already recognised the ironies and ills of the green revolution and wrote about them soon after his field trips. 'The new agricultural policy which has generated growth and prosperity is also the indirect cause of the widening of the gap between the rich and the poor' (1969b, p. 13). Big statistical surveys might have taken years to grind through their long agonies and come to the same conclusions. In 1969, though, enough was knowable and known for major policy conclusions to be drawn. Ladejinsky was not alone in having these insights. What is clear, though, is that he had exceptional experience and skill as a rural development tourist, and used an avenue of publication that was quick and influential. The problem is how to create Ladejinskys all over the world, together with equivalents of the *Economic and Political Weekly*, and to give them the freedom and encouragement to write and publish.

ii) Senaratne's windows into regions

S. P. F. Senaratne, as a social anthropologist, developed in Sri Lanka a method for bridging the gaps between economists and anthropologists, between practitioner-planners and academics, and between the macro-level of planning and the micro-level of village reality. Working first for the National Museum, then with a unit in the Ministry of Planning, and then with the Marga Institute (the Sri Lanka Centre for Development Studies), he initiated and managed a programme with ten villages chosen to represent conditions in ten regions of Sri Lanka. Each was studied by a graduate, and a body of comparable knowledge and understanding of each village was built up. The ten villages were expected to serve as 'windows into their respective regions' (Senaratne, 1978, p. 5). The researchers and their villages could then be used to respond to problems raised by planners, concerning for example 'the ineffectiveness of institutions, the failure of incentives, and the unpredictability of peasant response to urban logic' (*ibid*, p. 4).

Senaratne and his team performed two sorts of functions. One was to carry out investigations of specific problems, often before planners themselves really knew what issues they wanted the research to resolve. This proved difficult because the researchers were themselves expected to identify policy issues, formulate problems, provide answers and work out policy. The second function was to react *ad hoc* to the difficulties of various agencies and organisations and help them to identify societal factors which they might have neglected. This included brainstorming sessions, contributions to the planning of schemes, assessing the viability of schemes already planned, and most often diagnosing the causes for failures of schemes which had already been implemented. Despite a major difficulty of continuity among the researchers, this second function, Senaratne records, was by and large accomplished satisfactorily, contributing to an increase in perception and understanding by providing quite often 'a much needed correction in the form of a micro-perspective' (1978, p. 9).

Speed was often of the essence. When the project was in its early days, Senaratne wrote of planners that: 'An investigation which was likely to take two years was of little use in terms of their needs. Three months was the maximum but what was really appreciated was if a paper could be prepared within a fortnight' (1976, p. 3). Two years later, on the basis one assumes of hard experience, he shortened the period. Planners often wanted answers without any delay – 'off the cuff, within a day or two, a week at the outside' (1978, p. 4). It seems that an experienced team, with detailed knowledge of micro-environments, was able to respond usefully, each person speaking from knowledge of his or her own village.

iii) *Reconnaissance for crop improvement*

The 1970s witnessed a quiet revolution in the methods for appraising farming systems and identifying needs and opportunities for crop improvement: from the mid-1960s, careful and exhaustive surveys, such as those pioneered by David Norman and his colleagues at Ahmadu Bello University in Northern Nigeria, established and documented the complexity and logic of small farming activities under conditions of uncertainty. The approach was thorough and meticulous but the surveys expensive in time and resources. Some of the resulting wisdom of the mid-1970s was captured in a conference on field data collection in the social sciences (Kearl, 1976) held in 1974. The conference publication devoted over 12 per cent of its space to 'Considerations in Sampling', but less than 4 per cent to 'Preliminary Steps: area familiarization and reconnaissance for baseline surveys'. But the importance of the preliminaries to formal surveys was stressed, presaging the shift in the later 1970s to paying more attention to reconnaissance and exploratory surveys, and to making rapid methods of appraisal more reliable (*Agricultural Administration*, 1981; IDS, 1981; Pacey, 1981).

There were parallel developments at several of the international agricultural centres. In Eastern Africa, Michael Collinson of the International Maize and Wheat Improvement Centre (CIMMYT) pioneered an approach which sought to condense a wealth of insight and experience into a replicable field method (for a full account see Collinson, 1981). In this there are four phases: zoning – grouping farmers into relatively homogeneous populations by the present farming system, these populations being termed recommendation domains; evaluating local circumstances; rapid description and appraisal of the farming system (also described as an exploratory survey or pre-survey); and a verification survey. The rapid description and appraisal are pivotal. The researchers have an interview guide, divided into ten major sections which are, for both researchers and farmers, manageable 'bite-size chunks' (1981, p. 440). The interview guide is not a questionnaire, but a checklist. Not all of it is covered with any one farmer, or at any one sitting; but over a period each part of it is covered several times. As a result a farming system scenario can be written up, indicating problems and opportunities for crop improvement.[4]

One feature of these new methods has been communication and learning not only between farmers and researchers, but also between researchers of different disciplines. Agricultural economists working for CIMMYT have come to recommend a team of two, usually a plant-breeder or agronomist and an economist, spending one to three weeks on an exploratory survey. Another approach has been developed by Peter Hildebrand, working for ICTA in Guatemala. He has developed an ingenious and quick 'technology generating system', a key part of which is a week spent in the field by a team usually consisting of five social scientists (among whom there may be anthropologists, sociologists, economists or agricultural economists) who are paired each day with five agricultural scientists (among whom there may be both plant and animal technicians in entomology, breeding, pathology, physiology, etc.). Over five days, they change partners each day to reduce interviewer bias and to increase cross-disciplinary interchange. The group also meets each night to discuss the day's findings, make preliminary interpretations, and modify the investigation if necessary. At the end of a week, the many three-cornered discussions – between farmer, social scientist, and agricultural scientist – have produced proposals for improved farm practices. One may conjecture that knowing that these proposals are to be tried out with farmers during the next season encourages researchers to listen to and learn from farmers during the reconnaissance (for a fuller account see Hildebrand, 1981).

The approaches of Collinson and Hildebrand have in common a procedure which forces or precipitates learning from others – from farmers, and from other disciplines – in a manner which is systematic but open. Both depart from conventional ideas of rigour. And both concentrate on and expand the early stages of learning, maintaining flexibility and options much longer than previous approaches. In this respect they resemble the evolution of the human embryo, prolonging the early stages of development in order to increase competence. The question then arises whether a formal survey, like Collinson's verification survey, is needed at all. After several major exercises with the method, Collinson has written:

The Verification Survey represents the major commitment of professional time and funds in the sequence. So far this formal sampling of the population has always verified the findings of the pre-survey. There have been no major contradictions. It may be that even this low cost, single visit formal survey is superfluous so long as the Exploratory Survey is rigorous. At present the numbers which this formal survey provides are the only hard evidence produced by the diagnostic process. This is extremely important in convincing 'the Establishment' that there is a need for an understanding of small farmers as a prerequisite to *relevant* research and development efforts.

(1981, p. 444 his emphasis)

Once 'the Establishment' is convinced of the general principle, the method might, it seems, become even more cost-effective if it could abandon the verification survey. But in any case, with both Collinson's and Hildebrand's approaches, being inventive with the preliminaries of appraisal has paid off handsomely. Their methods enable researchers to learn quickly and cheaply from farmers and from each other and to move in a short time into relevant and promising experimental programmes with farmers.

iv) *BRAC and the net*

In the final example the poor themselves took part. The Bangladesh Rural Advancement Committee (BRAC), a non-government organisation, became concerned in 1979 about the effects of a drought and mounted an emergency relief programme in three areas. The strategy was to organise groups of landless people in each village, obtain access to land, and encourage them to start collective agricultural activities with their own resources initially, supported later by food for work. It became clear to BRAC fieldworkers that quite large-scale government relief operations were going on which, if successful, would have made their work largely unnecessary; but much of the relief was being intercepted by a small number of powerful men, well connected with government officers, who were a net between the landless and the central government. In order to get more resources through, it was not enough for the landless to organise; it was also essential to understand the system clearly and put pressure on the weakest points.

The research methods were simple. Those who developed

them considered that they could be repeated by any field worker who could read, write and do simple arithmetic. The main sources of information were the landless people of each village, but information was also obtained from the local elite themselves and from government officers. Ten villages were included. Incidents of exploitation and abuse were recorded. Then, after checking all the details with at least four separate sources, and sometimes up to fifty, the BRAC fieldworkers plotted all the connections involved in each incident to build up a picture of the network involved. Profiles of powerful people and their followers were compiled. Government machinery was contacted at all levels up to the district level to find out what was supposed to be done, in order to compare it with what was actually being done.

The outcome was the remarkable report, *The Net: Power Structure in Ten Villages* (BRAC, 1980), which describes those who are powerful and how they operate. The report penetrates exploitative activities which even social anthropologists have rarely revealed in this sort of detail. The concern began with food, but spread to include land, capital, the forest, education and law and order. The area was on the border with India and had a history of disturbance and refugeedom, with a population of Adivasis (tribal people) who are especially weak, and an Army presence. The conditions there may thus have been extreme. The effect is like the lancing of a long-festering abscess.

The fieldwork for this research covered only five months. The research was secondary to the main tasks of the fieldworkers. Yet the involvement of the landless themselves yielded a wealth and detail of information which taught outsiders much. Moreover, the landless participants gained in critical awareness. They had previously, individually, been aware of parts of the net. Through their joint research with outsiders they came to see how the parts fitted together, and also to realise that their strength as landless people lay in their unity and collective action.

Conclusions

These four approaches appear cost-effective in generating insights and action to improve the livelihoods of the poor, whether directly or through influencing opinion, policy and other research. Ladejinsky's two articles were published and republished, and widely read and quoted. Senaratne's windows into regions, quite apart from their immediate usefulness in Sri Lanka, have influenced approaches and experiments elsewhere. The work of Collinson, Hildebrand and others in crop improvement reconnaissance has changed priorities in agricultural research and has been disseminated internationally through the international agricultural centres, through conferences of agricultural economists and others, and within countries. The work of BRAC is too recent for more than a hope that quite apart from its purely local effects, it will be widely read and influential. If such research is not repeated in other places, it will be a sad reflection on what Gunnar Myrdal has so trenchantly condemned as the 'diplomacy in research' (1968, p. 12ff.).

Let us then see what these four approaches and experiences have in common.

All are *eclectic* and *inventive*. None involves either the extensive questionnaire survey or prolonged total immersion, though Senaratne's windows may combine elements of both of these. Ladejinsky picks up information wherever he can; he notices things and goes straight to talk to people, like the owner of the Massey-Ferguson tractor he saw ploughing a field next to the Purnea airstrip where he landed in Bihar. Collinson adapts the questionnaire method into bite-size chunks of a checklist, Hildebrand builds an interdisciplinary approach into the social relations of investigation. BRAC uses many informants, the philosophy of participatory research, and elements of network analysis and of the critical incident methods of management consultants. All reverse the tendency for the method to determine the problem; they allow the problem – the green revolution, planners' needs for insight into the micro-level, agricultural researchers' needs for priorities, and the exploitation of the weak and poor – to determine the choice and invention of the method.

All the approaches are, by the same token, *adaptable* and *open to information* other than that which is directly sought. All avoid the narrow vision of survey slavery and the costly inclusiveness of some total immersion. All have a focus but are able to expand it. Ladejinsky, though concerned with agricultural technology, concluded with observations about 'the social, religious, economic and political forms which govern the village' (1969b, p. 12). Senaratne's windows could be used to see many facets of the ten villages, once the initial data base had been built up. Collinson's approach allows scope for interviews to go where respondents wish, and in the BRAC research, the landless were free to say what they wished. This does not mean that any of these researchers were without preconceptions. But they were open rather than closed-minded about what they would notice and consider relevant.

All *spanned the two cultures*, addressing both practical problems and the academic world. The BRAC research was

71

almost a side effort, carried out 'simultaneously with more urgent work' (1980, p. 6). The link with practice and policy is reflected in each case in the timeliness of the report. We find here none of the long gestation periods of massive surveys or social anthropological treatises. Ladejinsky's articles were in print a few months after his field trips: the interval was five months for his Bihar visit, and one month from the second of his two visits to Punjab. Senaratne could respond to requests in a matter of days. Collinson and Hildebrand shortened the time taken between field investigation and the start of new agricultural research, or changes in research priorities; and BRAC carried out their fieldwork in five months, publishing *The Net* only three months later.

Finally, all made use of experience. Ladejinsky had a good sense of what he was looking for. Senaratne and his researchers, by virtue of the experience and knowledge derived from their fieldwork, were able to provide judgements at short notice. Collinson's long experience with small farming in Eastern Africa was transferred to the checklist which organised the coverage of the range of relevant questions. And BRAC, in a different way, through participatory research, mobilised the experience of the landless themselves.

In reversing conventions, the BRAC research went furthest. The other three approaches were concerned primarily with influencing the existing structure of power from above – the 'policy-makers', whether concerned with agricultural research or extension, or with any of a range of issues of planning. The BRAC research was also concerned with action at the local level:

> The first stage was to record carefully all the examples of oppressive, exploitative and illegal activities we could find. We did not have to go out to look for them very much, the landless and poor, who were the principal victims, came to us and as our study continued, their interest and analytical capacity increased to the point where they gave us pens and paper and insisted that we record everything. Of course none of the incidents we have recorded are new to them, they know about those things far better than us. But by linking incidents and activities from different villages, by comparing what happens with what is supposed to happen and above all by discussing and recording the oppressive activities of the powerful, as if they could be fully understood and then checked, we helped the landless to develop a new consciousness and militancy. Already they have started taking collective actions on certain issues and achieving limited success. From this point of view our investigation felt

like shining a torch into a dark room. Previously everyone knew some of the things that were going on because they were right in front of him, but it was in a shadowy, partial way. By adding his knowledge to that of others and then by analysing and calculating everything, they could see in a clear open way for the first time, and so realistically consider the possibility of change.

(1980, pp. 3–4)

The most obvious impact of the BRAC research may be local and direct, as the researchers were already seeing by the time they left. But much more significant for the conditions of the poor will be the way their report works its way through much longer causal chains in the cores of centralised knowledge and policy. Paradoxically, the report was possible because the researchers started not with research, but with the problems and knowledge of the landless, working with them on solutions. We have moved a long way in the research approach, from participant observation to participant organisation. Purists may throw up their hands in horror and point to the danger of distortion and propaganda. But in the next decade those outsiders who have the courage and vision for such reversals, and who communicate their experience widely to others, will be at one cutting edge of rural research.

As with this BRAC example, some of the most exciting and useful work does not fit common categories. It is neither purely 'research' in the observer-observed and data-collecting senses, nor purely 'action' in the sense of outsiders acting on rural conditions and people, nor purely 'consciencisation' in the sense of enabling those who are deprived to become more aware of their conditions and capabilities and so more able to choose and act themselves. It is, rather, mixtures of these. The work and writings of Paulo Freire (e.g. 1970) whose pedagogy of the oppressed enables the poor to look critically at their world, to break out of their 'culture of silence', and to take control of their own destinies, has been an inspiration to those who have been seeking methods of research in which rural people are actors rather than objects of observation and sources of data (e.g. Haque, Mehta, Rahman and Wignaraja, 1977). 'Participatory research' describes methods in which rural people and outsiders are partners. One good aspect of this new work is respect for the poor. Another is greater sensitivity to the dangers in traditional research of exploitative data-mining, taking the time of busy poor people and giving little or nothing back.

But work on this frontier where research, action and consciencisation overlap should also be looked at critically.

Activism by researchers and research by activists are vulnerable to sudden interruption and do not combine well with the collection of data according to a routine, where this is necessary. How good such activist research is depends, as with all research and action, on the purpose, the costs, the alternatives, and replicability and impact. The impact of research and action with and by the poor will be slight if it changes only one small microcosm at the periphery; it will be more cost-effective if it spreads laterally or if it links back with and affects the cores of knowledge and power.

Finally, the conclusion from this discussion is that conventional and professionally respectable methods for rural research are often inefficient. The search is for approaches which are open to the unexpected, and able to see into, and out from, the predicament of the rural poor themselves. For the future, three poles of concentration may serve well: first, long-term, careful investigation, including statistical analysis, and involving social, medical and natural scientists; second, *ad hoc*, inventive work, improvising and adapting for the sake of timeliness and cost-effectiveness; and third, sensitive research which shifts initiative to rural people as partners in learning, enabling them to use and augment their own skills, knowledge and power.

Notes

1 Not entirely, however. See, for example, Kearl, 1976.
2 For a critique, see Cassen, 1976, pp. 793–795.
3 But for an excellent example see Gulati, 1981.
4 This summary does not do justice to the method. The reader is referred to CIMMYT, 1977a, 1977b, and 1978, and Collinson, 1981, for accounts which describe it in more detail as it evolved. See also CIMMYT, 1980 (part of which is summarised in IADS, 1981) for a guide to collaborative research by biologists and economists.

CHAPTER FOUR

Whose knowledge?

> The development profession suffers from an entrenched superiority complex with respect to the small farmer. We believe our modern technology is infinitely superior to his. We conduct our research and assistance efforts as if we knew everything and our clients nothing.
>
> Hatch, 1976, pp. 6–7

> In practice, the comparison with knowledge of western scientists is rendered . . . difficult . . . since the !Kung appear to know a good deal more about many subjects than do the scientists.
>
> Blurton Jones and Konner, 1976, p. 328

> Mwalimu Nyerere is right. So-called leaders do entirely too much talking to the peasants. No one ever wants to listen to them.
>
> A Tanzanian agricultural extension worker (Thomas, 1977, p. 30)

The links of modern scientific knowledge with wealth, power and prestige condition outsiders to despise and ignore rural people's own knowledge. Priorities in crop, livestock and forestry research reflect biases against what matters to poor rural people. Rural people's knowledge is often superior to that of outsiders. Examples can be found in mixed cropping, knowledge of the environment, abilities to observe and discriminate, and results of rural people's experiments. Rural people's knowledge and modern scientific knowledge are complementary in their strengths and weaknesses. Combined they may achieve what neither would alone. For such combinations, outsider professionals have to step down off their pedestals, and sit down, listen and learn.

Knowledge, power and prejudice

It is a truism that knowledge is power. At the crudest level, technological 'superiority' carries superior physical power:

> Whatever happens we have got
> the Maxim gun and they have not

But the relationship has wider and subtler ramifications. Those who are powerful and dominant have the greatest accumulations of wealth, a centralised and interconnected system of communication, an ability to determine what new knowledge shall be created, and control over flows of information from the centre to the rural periphery. The association of outsiders' modern scientific knowledge with wealth, power and prestige generates and sustains beliefs in its universal superiority, indeed beliefs that it is the only knowledge of any significance. After all, it is this knowledge which has made possible the cities, roads, railways, telephones, transistors, schools, hospitals, medicines and guns which have penetrated and transformed many rural areas. Uneducated rural people see that this sort of knowledge, acquired through schooling, leads upward and away from rural life to urban opportunities and rewards.

Those who acquire formal education and training then have a personal stake in the system. If they live and work in rural areas they derive their status partly from their positions as bearers of modern knowledge. School teachers, health workers, agricultural extension staff, and other rural officials look upwards and towards the centre for authority and enlightenment. They, like others with formal education and training, need to believe that the knowledge and skills they have acquired are superior and that uneducated and untrained rural people are ignorant and unskilled. From rich-country professionals and urban-based professionals in third world countries right down to the lowliest extension workers it is a common assumption that the modern scientific knowledge of the centre is sophisticated, advanced and valid and, conversely, that whatever rural people may know will be unsystematic, imprecise, superficial and often plain wrong. Development then entails disseminating this modern, scientific, and sophisticated knowledge to inform and uplift the rural masses. Knowledge flows in one direction only – downwards – from those who are strong, educated and enlightened, towards those who are weak, ignorant and in darkness.

Outsiders' biases

In rural development, the centre-periphery biases of outsiders' knowledge are reflected in the concentration of research, publication, training and extension on what is exotic rather than indigenous, mechanical rather than human, chemical rather than organic, and marketed rather than consumed. It is reinforced by other biases – towards what concerns men rather than women, adults rather than children, the clean rather than the dirty, and, pervasively, the rich rather than the poor.[1] Some of these points can be illustrated from research and extension in the three domains of crops, livestock, and forestry.

In crop research, priority, prestige and promotion have gone with work on crops for export, grown usually by plantations, large farmers, the better-off small farmers, and the men of the household rather than the women. These crops include rubber, tea, sisal, jute, palm oil, cotton, coffee and cocoa. Since the 1970s and the initiatives of the international agricultural centres, more attention than before is being paid to poor people's and women's crops for subsistence – such as the millets, sorghum, cowpeas, chickpeas, cassava (tapioca, manioc, yucca), sweet potatoes, and yams. But they are still often overlooked. Sometimes they do not even appear in agricultural production statistics, as with cassava in Zambia, although cassava is grown by over half the Zambian rural population and is for many of them the basic staple, and for most the fall-back food of last resort (ILO, 1981, p. 59). In Zambia, too, in 1980, there was only one solitary research agronomist working on cassava. Sometimes, also, research on crops which poor people eat is geared not to what people need (usually more calories) but to what makes large-scale livestock enterprises profitable (more protein), as with sorghum breeding for animal forage rather than human consumption in north-east Brazil (Sanders, 1980). And following on from priorities in research, so agricultural extension for small farmers has concentrated on cash crops and those, usually the better-off minority of farmers, who are able to grow them.[2]

Livestock research and extension follow similar patterns.[3] The carriers of modernity have often been exotic cattle, bred for and suitable for temperate climates, and needing special care and coddling to survive in the tropics. There is here something of a professional fixation. It was at one time fashionable to believe that certain East African tribes had an irrational, emotional and aesthetic attachment to cattle, dubbed the 'cattle complex'. But it was veterinary and animal husbandry experts who suffered most from this complaint. Their attachment to exotic cattle to the exclusion of native beasts and other domestic species may have had aesthetic and emotional dimensions, but there was also a degree of irrationality. To be sure, there were successes, as with small farming dairy cattle in Kenya. But more generally, what Robert McDowell has called the 'milk and meat complex' of expatriates was based on their professional training for the conditions and needs of rich, temperate countries and was inappropriate for those which were poor and tropical.

For the poorer rural people, exotic cattle are usually either impossible or unattractive as investments. In economic terms they are 'lumpy': they come in large units of value which are not divisible while alive and which do not store well when dead. This concentrates risks. Moreover, they are vulnerable to tick-borne and other tropical diseases (rinderpest, foot-and-mouth, East Coast Fever . . .). Only households who are already well buffered against contingencies may be sensible to risk exotic cattle. In contrast, the animals of the rural poor are cheaper and smaller: either physically less large native cattle, partly resistant to local diseases, or other usually smaller animals – donkeys, mules, yaks, llamas, pigs, sheep, goats, turkeys, hens, guinea fowl, pigeons, ducks, rabbits, and guinea pigs. Of these, goats[4] and donkeys have been especially neglected.

Goats have many advantages for poor people: they are less lumpy than cattle and so spread risks better; it is cheaper and easier to obtain a few to breed up; they reproduce fast; they can be used as buffers to raise cash for small or urgent needs without selling a major asset; they are a larder of food that can be used at any time; they can be herded by children; they browse on bushes and can be managed to produce milk in the dry season when milk from cattle drops off (Swift, 1981a, pp. 82–84); and they can be used for entertaining visitors or for special occasions and feasts. Yet they have been the subject of relatively little research, are ignored in most government extension programmes, and are regarded by some professionals as a pest. Even the dung of goats has been neglected in work on biogas in India although, as Amulya Reddy has pointed out, it provides a way for some of the poor who do not own cattle but who do have goats to contribute to a communal digester.

Donkeys, even worse, are a joke and their value little recognised. At a subliminal level, those who have had an English or French Language education may despise donkeys partly through the associations of ass and âne applied to human stupidity. The Shorter Oxford English Dictionary (1955) has for donkey: 'A stupid or silly person', and 'An ignorant fellow, a conceited dolt'. Under 'goat' we find 'to play or act the (giddy) goat' meaning to play the fool. Mules, however, do a little better. The SOED, while conceding 'A stupid or obstinate person' is gracious enough to assert, though in brackets, that '(Without good grounds, the mule is a proverbial type of obstinacy)' (my emphasis). The stereotype of the despised donkey is reflected in G. K. Chesterton's verse:

> When fishes flew and forests walked
> and figs grew upon thorn.
> Some moment when the moon was blood
> Then surely I was born.
>
> With monstrous head and sickening cry
> and ears like errant wings,
> The devil's walking parody
> Of all four-footed things.

But it concludes:

> Fools! For I also had my hour;
> One far fierce hour and sweet:
> There was a shout about my ears,
> and palms before my feet.

Perhaps the donkey's hour will come in professional research; or perhaps since it is so tough a beast, so well adapted to bad conditions, it has already achieved a sort of perfection beyond the power of research to improve. This is consistent with Polly Hill's praise of them in Hausaland: 'These small, sturdy, tax-free beasts can manage loads of 200 lbs upwards . . .' (1972, p. 227). But she notes that they have been overlooked. 'Although donkeys are the local camels of Hausaland, and are a most valuable source of manure, their importance has been neglected in literature – for instance by (an FAO report) which regards headloading as the only alternative to road, rail and water transport' (*ibid*, p. 226). Donkeys are important for the earning capacity of the poorer rural people. Mules too can play an important part in the economy of a poor family; the distress sale of a mule by one of the five families studied by Oscar Lewis in Mexico meant less wood to sell, less earned by carting crops, and more trips to bring in the family's own crops (Lewis, 1959, pp. 39–40, and pp. 118–19 below).

No Nobel prizes have been awarded for work on donkeys, goats or mules.

Forestry has similar biases in research and extension. Tropical forestry has paid much attention to the introduction of exotic trees and their culture in single-species stands in plantations. Little attention has been paid to indigenous species and their cultivation. In their study of the knowledge of vegetation of the Mbeere in Kenya, Brokensha and Riley point out the lack of exotic species which provide good timber, while foresters were ignorant of indigenous species which did provide good timber (1980, pp. 122–3).

Another pervasive bias is against the technology and needs of rural women. Until recently, little attention was paid to home gardens and backyard farming, often sources of small but vital incomes for women. Domestic technology – for processing food, cooking, cleaning, sewing, fetching firewood, carrying water – all traditional responsibilities of rural women, is regarded as uninteresting, a low priority. When the person-hours devoted to these activities are considered, and the drudgery they entail, it is a grave reflection on those with power how miniscule has been the attempt to improve the technology of such activities. The processing of staple foods (cassava, millets, sorghum, paddy) by hand is a gruelling task for hundreds of millions of women, yet easier domestic processing is little recognised as a criterion in seed-breeding, and few engineers or scientists have turned their minds and energies to seeing how, from the woman's point of view, the process could be made easier.[5]

The pro-male and anti-female bias applies in other spheres too. Ploughing, mainly carried out by men, has received more attention than weeding or transplanting, mainly carried out by women. Cash crops, from which male heads of household benefit disproportionately, have received more research attention than subsistence crops, which are more the concern of women. Even now, after a massive shift of rhetoric and a notably less massive shift of real priorities towards rural women and their needs, not much more than a modest foothold has been established in the field of technical scientific research and government extension. There is a male cognitive problem. To take but one example – 'Foresters in Senegal say repeatedly that women cannot be involved in projects as they do not and cannot plant trees, when Senegalese women have traditionally raised crops as well as planted trees in the courtyards of these foresters' own homes' (Hoskins, 1979, p. 14). As a first step from this stage of denial, tokenism is becoming more common – the appointment of a woman staff member, or of the setting up of a (small, weak) extension section for women for what in the UN's language, are described as 'optical' or more generally as 'cosmetic' purposes. It is rare indeed to find substantial changes in perception, attitude or behaviour among the male majority of professionals. Scientific and engineering establishments in particular remain heavily male-dominated and are usually still a very long way from recognising, let alone giving balanced attention to, the needs of rural women.

Other prejudices also make it hard for professional outsiders to perceive what is important to poor rural people or its advantages. The rich despise the things of the poor; the powerful despise the things of the weak; the learned despise the things of those they think ignorant. In the simple and eloquent, if urban and male-biased, words of Ecclesiastes:

> There was a little city, and few men within it; and there came a great king against it, and besieged it, and built great bulwarks against it;
>
> Now there was found in it a poor wise man, and he by his wisdom delivered the city; yet no man remembered that same poor man. Then said I, Wisdom is better than strength: nevertheless the poor man's wisdom is despised, and his words are not heard.
>
> The Bible, Authorised Version, Ecclesiastes 9, verses 14–16

In the Philippines, Noel D. Vietmayer reports that a visitor discussed the winged bean with an influential Filipino family. They were incredulous that such a miraculous plant could exist. So on a hunch the visitor took them out to the back to the servants' quarters. There climbing along a fence was a winged bean plant laden with pods. '"But that's just *sequidillas*," they said, disappointment echoing in their voices. "It's only a poor man's crop".' (Goering, 1979, p. 1).

In Kenya, the *mukau* tree has long been recognised by the Mbeere people as a valuable resource, pre-eminent among local trees; it produces a bole that can be longitudinally split for house construction poles which are relatively straight, have an untwisted grain, warp less than other woods and are moderately durable in the ground. Brokensha and Riley consider that this is probably the only indigenous timber tree that has been deliberately encouraged and conserved on a wide scale. The seedlings 'which appear to germinate successfully once the seeds of the fruits browsed by goats have been passed in their droppings, are, when found in clearing land for cultivation, protected and reared as individual property'. They continue:

> At the time of our enquiries, few forestry officials knew that this valuable tree could be germinated by people: indeed, some authorities denied this verbally and in print. Yet as we were told rather scathingly by an old man, who was surprised by our ignorance 'every uncircumcised herd-boy knows how

we germinate mukau' (Brokensha and Riley, 1980, p. 123).

Perhaps the cognitive problems of professionals are most intractable when several biases interlock. The germination of the mukau tree combines the low status of the indigenous trees, the low status of the goat, and the even lower status of goats' droppings. It is no wonder that they pass unnoticed and their potential is overlooked: a programme to germinate mukau trees, however admirable a long-term investment for the people of Mbeere, would require foresters to collect the seeds of an indigenous tree, keep goats, feed the seeds to the goats, collect the goats' droppings, and then tend the droppings with care.

Rural people's knowledge

Centralised urban and professional power, knowledge and values have flowed out over and often failed to recognise the knowledge of rural people themselves. An exception has been social anthropologists who have been at pains to experience cultures other than their own from inside, and to learn and understand the values and knowledge of those cultures. The result has been recognition of the complexity, variety and validity of indigenous knowledge systems.[6] These have been variously described as people's science, ethnoscience, folk-ecology, and village science (see e.g. Barker et al., 1977, pp. 2–3). The ethno prefix is widely used, as in ethno-ecology, ethno-soil science, ethno-agronomy, ethno-anatomy, ethno-taxonomy, ethno-botany, ethno-medicine, ethno-linguistics and ethno-aesthetics. Others have written about indigenous technical knowledge (ITK) (IDS, 1979) which can be contrasted with modern scientific knowledge (MSK). More simply 'local knowledge' has also been used (Korten and Uphoff, 1981).

There are problems with all these terms.

People's science can be used to describe the knowledge system of a group of rural people. But in addition, it has been used to describe not the science of the people, in this sense, but science for the people, making the knowledge of formal science available to them. Thus, describing the People's Science Movement in Kerala, K. P. Kannan (1981) mentions 'study classes and lectures wherein the received knowledge in science and technology was shared with the people'. These science classes 'dispelled any doubts about people's ability to think in ways and methods other than the ones they were accustomed to'. There was a dialogue with rural people, but, it seems, no attempt to learn from them.

Ethnoscience also has a range of meanings. It can be used to describe indigenous knowledge systems themselves. But to one writer, this includes Western science as one among many ethnosciences (Meehan, 1980, p. 385). For some others, the term ethnoscience also refers to the methodology for eliciting, translating and interpreting the knowledge system of a particular culture (Werner and Begishe, 1980, pp. 151–2).

Indigenous implies originating from and naturally produced in an area, but rural people's knowledge is also added to, influenced by, and destroyed by knowledge from outside the area. However, the literature on indigenous knowledge systems (e.g. Brokensha, Warren and Werner, 1980) and on indigenous technical knowledge (ITK) (IDS, 1979) has served to emphasise the separateness, sophistication and validity of the knowledge of groups of rural people, and the 'technical' in ITK also has a healthy effect in emphasising the practical nature of much of this knowledge.

Local knowledge is tempting for its simplicity. Local knowledge of rural peripheries can be contrasted with centralised knowledge of urban cores. But a weakness is the commonsense interpretation that it refers to knowledge of a local environment, rather than to the knowledge of people existing as a system of concepts, beliefs, and ways of learning.

While using some of these terms where appropriate, I shall adopt *rural people's knowledge* as my most inclusive term. The 'rural' includes those farmers, both small and large, who are thoroughly in the market, purchasing inputs and selling cash crops, as well as groups like the San of the Kalahari or the Hanunóo swidden cultivators of the Philippines who have been much more autonomous. The 'people's' part of the term emphasises that much of the knowledge is located in people and only rarely written down. 'Knowledge' refers to the whole system of knowledge, including concepts, beliefs and perceptions, the stock of knowledge, and the processes whereby it is acquired, augmented, stored, and transmitted.

Outsiders are hindered from appreciating and learning from rural people's knowledge by many forces. Besides power, professionalism, prestige, lack of contact, problems of language, and sheer prejudice, another factor is the gap between practitioner and academic cultures. Those outsiders who have most studied and understood rural people's knowledge have been social anthropologists: they have concentrated somewhat on remote and isolated people who often prove to have rich and complex indigenous knowledge systems. The painstaking studies of these social anthropologists have led them off into what, to the uninitiated, must appear a morass of detail, and into what for practitioners is bound to appear an esoteric analysis of the cognitive systems of particular rural cultures. To the practitioner's question – so what? – they may not always be ready with an answer. Some academics may even be open to charges of romanticism, and stamp-collecting.

The result of this communication gap between academics and practitioners is unfortunate. On the one hand we have rural people and a handful of researchers with access to and understanding of rich and detailed systems of knowledge which do not influence development; and on the other we have government organisations and staff engaged in development but ignorant of and conditioned to despise that knowledge. To bridge the gap requires reversals to offset the grotesquely unequal balance between outsiders' knowledge and rural people's knowledge. Outsiders' knowledge (modern, scientific . . .) is accessible to them in books, and information retrieval systems, is easily communicated, and is taught all over the world. It both supports the state and the state apparatus and is supported and propagated by it. It can claim to be universal. In contrast, the knowledge of any group of rural people is accessible to outsiders only through learning from rural people themselves or, rarely, through ethnographic literature coded in anthropological jargon. Moreover, rural people's knowledge exists in innumerable forms among innumerable groups of people in innumerable environments. Outsiders, so well connected to centralised knowledge, have access in the written form to only a minute proportion of that of rural people. It is the powerful who are ignorant. It is they who have to begin as learners, and rural people who can instruct them.

In seeking a balanced view of rural people's knowledge, it is as well to note that it can be overvalued as well as despised. The social anthropologists who have elicited it have been well-disposed towards it and have praised its richness and interest. However, the communities which are most cited tend to be those which have revealed the most remarkable knowledge; in the discussion which follows, this applies to the Hanunóo swidden cultivators of the Philippines studied by Conklin (1957, 1969) and the San (Bushmen) of the Kalahari studied by several authorities. The spread and depth of knowledge in a community may also be exaggerated through the selection of informants. Any prudent investigator will seek out those who know most:

Generally, the best information about the small annual herbs is obtained from older women; herd-boys, being always hungry and also experimental, are experts on the range of wild edible fruits; honey-collectors show the most detailed knowledge of flowering sequences, and indeed know most differential characteristics of their local plants. Yet even within a group, one individual will stand out because of keen powers of observation, prodigious memory, curiosity and intellect.

(Brokensha and Riley, 1980, p. 121)

What is recorded may then represent what those who are best informed know, and is by no means an average for the society as a whole. Nor is rural people's knowledge always valid or useful. A further danger is that some observers may be tempted to revive the Noble Savage, or to reincarnate him[7] as the Rational Peasant whose actions are perfectly judged exercises in optimisation that even well-informed computers can only struggle to simulate. But these positive biases may be no bad thing. The colonising force of outsiders' knowledge is programmed to override and bury other paradigms and to impose its own. It needs to be offset by countervailing power. To balance it not only requires an independent and open mind; it also requires positive discrimination.

Rural people's knowledge, and especially indigenous knowledge systems, have many dimensions, including linguistics, medicine, clinical psychology, botany, zoology, ethology, ecology, climate, agriculture, animal husbandry, and craft skills. Its validity and range have been neglected in all of these. In supporting reversals in the attitudes and behaviour of outsiders, four out of many aspects will be singled out for comment. These are farming practices; knowledge of the environment; rural people's faculties; and rural people's experiments.

Farming practices

Many of the practices of small farmers which were once regarded as primitive or misguided are now recognised as sophisticated and appropriate. Examples of this understanding include various forms of sparing tillage and shifting cultivation (Allan, 1965; Boserup, 1965; Harwood, 1979; de Schlippe, 1956). Only one such practice will be described here, by way of illustration: mixed cropping in West and East Africa. This refers to the growing of two or more crops simultaneously on the same land. Mixed cropping has been and remains a widespread technique in small farming in tropical Africa and elsewhere. Yet for many years it was regarded as backward. Since agricultural research was confined to pure stands of crops, it was only natural that the advice emanating from research stations and conveyed to farmers was also to plant pure stands. When small farmers continued to plant mixtures they were branded as primitive, conservative, ignorant, lazy and unprogressive.

With hindsight, the agricultural researchers and extension staff are easily condemned. But there were many reasons for their behaviour. Many of the researchers were foreigners with a background and training in the agriculture of temperate climates, with large farms and mechanised row-planting, weeding and harvesting, where pure stands made economic and agronomic sense. The agricultural development policies of the colonial countries where they worked aimed to increase the output of single crops, mostly cash crops for export but also food crops for domestic consumption. Monoculture was practised by large (plantation, European settler) farmers who influenced research policy. The organisation and rewards of agricultural research also pushed researchers towards work on only one crop at a time: crop-specific teams were, and still are, a simple way to organise research, and work on one crop at a time is statistically simpler than work on intercropping with two or more. On top of all this, in most of Africa, expatriates conducting agricultural research suffered from cultural conditioning which made it difficult for them to see indigenous farming as anything but backward. The model in their minds was a tidy, geometrical, mechanised field in Europe or North America. The higgledy-piggledy muddle (as it seemed) of mixed cropping on African farmers' fields scarcely appeared a place to learn anything.

And yet it was (Belshaw and Hall, 1972; Norman, 1974; Belshaw, 1979). Not only have many of the supposedly irrational and wasteful practices of traditional African farming been found to be prudent and sound, but mixed cropping has been shown to have many advantages, including:

- different rooting systems exploit different levels in the soil profile for moisture and nutrients;
- one crop may provide a favourable micro-climate for another;
- nitrogen-fixing plants fertilise non-nitrogen fixing plants;
- crops which are scattered among others are less vulnerable to pest attacks than single stands;
- labour requirements are less, especially in reducing weeds;
- labour peaks are spread out;
- more moisture is retained in the soil;
- returns are higher per unit of land;
- successive sowing of crop mixtures supplies a mixed diet over an extended harvesting period;

- risk is less;
- where labour is a constraint, the returns to labour are increased at the time of the year when labour is limiting.

Not all of these may apply in every case. But the list is impressive. So is the fact that it took organised agricultural research decades to realise that what appeared primitive and unprogressive was complex and sophisticated. Small farmers are, after all, professionals. They cannot afford not to be. And as professionals they have much to teach.

Knowledge of the environment

Knowledge of the observable environment is also often very detailed. There is a debate (Howes, 1979) as to what extent such knowledge is utilitarian, and to what extent it reflects innate curiosity and the spirit of enquiry. It may be significant that the knowledge which presents the most categories and the finest discrimination is that of communities which live in environments with much diversity (presenting the opportunity), and/or which live near the margins of survival (presenting the need). Ethnobotany provides striking examples. Heinz and Maguire record that an !Ko bushwoman who was considered to have an average knowledge of plant lore for an adult member of her community, could recognise, identify and name 206 out of 211 plants, in spite of the effects on the specimens of a severe drought; and they considered that at least 300 plants were part of the generalised botanical knowledge of a Bushman (n.d., p. 43). Even more remarkable Conklin noted for the Hanunóo swidden cultivators in the Philippines:

> More than 450 animal types and over 1600 plant types are distinguished ... of some 1500 'useful' plant types over 430 are cultigens (most of which are swidden-grown), existing only by virtue of the conscious domestication of the Hanunóo. Partly as a result of this intensified interest in plant domestication and detailed knowledge of minute differences in vegetative structures, Hanunóo plant categories outnumber, by more than 400 types, the taxonomic species into which the same local flora is grouped by systematic botanists.
>   (Conklin, 1969, pp. 229–230)

For the Andes, Brush reports 'vast numbers' of locally named varieties or cultivars of potatoes, with up to one hundred in any locality and several thousand in the central Andes alone (1980, p. 40). For any rural society, it is difficult to foresee how rich the naming of types will be but there may often be many more than the outsider is predisposed to expect.

Soils and land types are another domain where local knowledge is strongly based. Soil types are usually distinguished by colour and texture. Some farmers in Nigeria use colour to identify degrees of soil fertility (Netting, 1968). In one case in Malaysia soils are also discriminated into three categories by taste, as sweet, neutral and sour – categories which correlate significantly with pH levels (Weinstock, 1977). The Hanunóo are reported to have ten basic and thirty derivative soil and mineral categories, four different terms for describing the firmness of soil, nine colour categories to reflect its properties, five different topographical types, and three different ways of categorising slopes (Conklin, 1957, p. 36). Soil colour is used by Somalis in Northern Kenya to distinguish soil-vegetation associations (Chambers, 1969), the strongest distinction there, as in other parts of Africa, being between red or dark brown and black soils. In Bangladesh indigenous land classification is based on the depth of flooding and associated differences in cultivation practices: six different depths are distinguished. While depth of flooding is not the only land characteristic important to cropping, it is 'certainly one of the most important considerations for use of some areas in Bangladesh' (Brammer, 1977). Nor are rural people's categories for land and soil limited to the third world. Stephen Kraft at Cornell University is reported to have found farmers in up-state New York to have 18 operational categories for land types, including such aspects as drainage, rockiness, slope, and duration of frost, and to find these more useful than the USDA soil classifications.

Climate is another sphere in which local knowledge can be strong and local lore soundly based. W. Reed (1970) made an intriguing discovery through his study of pest insects in East Africa. He collected them at night by attracting them to a light. The effectiveness of this method varied with the amount of competing light from the moon. This led him to become interested in cloudiness in relation to phases of the moon. He found that farmers generally sowed according to the phase of the moon, believing that there were lunar phases in rainfall. The Meteorological Office in Nairobi was sceptical, but Reed analysed five and a half years of rainfall data which supported the belief. The Meteorological Office was not at first convinced, but a subsequent analysis of rainfall at 200 sites near Nairobi confirmed an association between rainfall and lunar phase. The scientific explanation is that the lunar phase influences the amount of dust entering the earth's atmosphere which seeds rainfall. In Reed's words, 'The experts claimed that such effects could not affect rainfall in the tropics but the local farmers knew better'.[8]

Examples might be multiplied to cover the seasons, water sources, animal behaviour, insects and other invertebrates, livestock and livestock husbandry, and micro-environments. Rural people discriminate these not only through categories and their indicators and boundaries, but also in terms of location and timing. People often know not only the *what*, but also the *where* and the *when* of plants, animals, water, fruits and other elements of their environment.

Rural people's faculties

A strength of rural people's knowledge is the faculties which maintain, extend, and correct it. These may include acute observation, good memory for detail, and transmission through teaching, apprenticeship, and story-telling. These are needed because of a high wastage and replacement rate, much faster than with outsiders' knowledge stored on paper, in libraries, and on computer tapes. Such rural knowledge is at the same time vulnerable and adaptable. It is continuously lost through death; it is continuously renewed and corrected through observation. The Gourma of Upper Volta have a saying '*A yaa nua, a ba bandi* – How can you know if you have not seen?' (Swanson, 1980, p. 82).

Much of the knowledge of animal behaviour of the !kung San of the Kalahari is based on direct observation and respect for evidence. Blurton Jones and Konner found in seminars with San that they distinguished sharply between hearsay and direct observation; that in one discussion 'there was a striking rejoinder by an elderly man that his colleagues should speak only if they have seen things happen' (1976, p. 330); a readiness to admit ignorance; a lack of defensiveness when asked how they knew something – such questions typically leading to long and careful descriptions of observations or of tracking evidence; and a delight in long and detailed story-telling, including mimicry. Blurton Jones and Konner concluded that !kung observational methods resemble those of modern-day western ethology in attention to detail, in distinguishing data from hearsay, and in general freedom from inference. 'In these respects their observations are superior to those of naturalists such as Gilbert White and Aristotle, and very sophisticated indeed when compared with the legions of animal behaviourists among western hunters, gamekeepers, and pet owners' (1976, p. 333). Those who have to survive in extreme conditions cannot afford inaccurate observations or misleading inferences. For other rural people in less extreme conditions, and more so for those with secure or affluent livelihoods, there is more leeway, and their knowledge may be correspondingly less sharp and exact.

Rural people's knowledge can be underpinned and enhanced by a richness of discrimination not easily available to outsiders' science. This derives from an ability to use a wider range of experience and more of the human senses than a scientist. Two examples can illustrate this. First, !ko San identifying plants do not rely merely on visual appearance, as a conventional scientific taxonomist might. Rather:

> The !ko Bushman after his initial scrutiny, will sometimes be observed to touch or feel by rubbing between his fingers certain plant parts. He may then carefully smell and also taste these parts. Finally he may repeat the whole of this procedure after he has crushed or teased apart the feature being examined. All these observations can be of great assistance to him in successfully establishing the identity of a plant which for various reasons may offer problems in straight-forward identification. It may be mentioned that this procedure is but rarely adopted by the conventional ecologist and for obvious reasons usually cannot be adopted by the indoor taxonomist. Thus again is made manifest the detailed and in fact superior approach of the !ko Bushman to one aspect of plant taxonomy, and the welter of facts which when coordinated and learned form the basis and substance of his practical plant lore.
>
> (Heinz and MacGuire, n.d., p. 13)

A second example has been presented by Graham Chapman (1977). He sought to compare Bihari folklore about climate with what had been observed scientifically. He could obtain records giving rainfall for each 24 hour period measured precisely to within 1 mm for six stations for the period 1891 to 1965 with very few gaps. But he had only occasional average data for longer time periods for temperature and wind. A further problem with the scientific data was that they were aggregated according to the European calendar, while the intervals of the Bihar calendar better fitted and described local seasonal changes. As Chapman points out, the folklore contains very coarsely observed data for several variables. But this may be superior to more precisely observed data on only one variable. In information theory, five variables each measured on a scale of only three categories (for example hot-warm-cold, or wet-damp-dry) can convey as much information as a single variable on a scale of 1 to 243. Rural people's knowledge can have here a major advantage over that of outsiders. Whether it is San seeing, touching, crushing, smelling and tasting different parts of a plant, or a Bihari sensing insolation, cloudiness, cloud shapes, humidity, temperature, weather-related animal behaviour, and rainfall, they can achieve a richness of observation and a fineness of discrimination which would only be accessible to organised science through a vast exercise of measurement and computing. Five senses, keen observation and a good memory go a long way.

Rural people's experiments

Perhaps the least recognised aspect of rural people's knowledge is its experimental nature. Michael Howes has postulated 'the likely universality of what might best be described as the "experimental mentality" – at least where relatively little risk is entailed' (1979, p. 18). Experiments can also be risk-minimising. When cassava came to Nigeria, it was known to be sometimes toxic; so to establish the conditions in which it could be eaten safely by humans, it was first fed to goats and dogs. Jeremy Swift reports that pastoralists in Mali noticed that drinking a lot of tea made people nervous and irritable and argued whether the tea or the sugar was the agent. To find out, they took the liver of a freshly slaughtered animal and poured on first, water and sugar, which gave no reaction; and then tea without sugar, which did give a reaction. It is scarcely surprising if an experimental mentality should be part of the human condition, at least among those whose education has not suffocated it; prudent curiosity and judicious testing have survival value.

The experimental approach is marked in agriculture. It is implicit in the selection of seeds or clones. Andean farmers select potato varieties according to several criteria (Brush, 1980, pp. 45–46). The Director of the Bangladesh Rice Research Institute knows of three cases where farmers have made their own selections from IR-8, one of the earliest high-yielding dwarf varieties of rice released by the International Rice Research Institute; in all cases they selected greater plant height for conditions where flooding was difficult to control (Brammer, 1980, p. 25). Small farmers in Kenya tend and nurture clones for their best tea bushes in improvised greenhouses (Fitzgerald, 1980).

None of this should occasion surprise. It is, after all, through such selection that domestic crops and livestock have evolved over the ages. But experimentation goes further than this. There is curiosity in trying out new plants and new methods of cultivation. The Hanunóo are reported to show great interest in unfamiliar plants which are tested on small plots near to people's homes (Conklin, 1957, p. 10). The adoption of any new crop or practice is an individual experiment. Farmers can also be ahead of scientists in breeding techniques. Paul Richards reports this anecdote. In Nigeria, a scientist made a breakthrough. Yam propagation is normally vegetative, but the scientist managed under his experimental conditions to breed some yams from seed, as he believed for the first time. However, on a chance encounter a farmer said that he had himself succeeded in doing this; and not only that, but he had also discovered, as had the scientist, that although the first generation of tubers were small, second and subsequent generations were of normal size. Legend concludes this anecdote with the scientist thanking God that farmers did not compete in writing scientific papers.

The readiness of small farmers to experiment and innovate on their own has been obscured by the preoccupation in the social sciences with the agricultural research, extension and communication which are carried out through official organisations. The fact is that innovations which farmers can manage and find are good spread very rapidly indeed, through innumerable personal trials. In Sri Lanka, the new rice variety H4, released for general cultivation in 1953 (Dias, 1977, p. 57), raised yields by some 50 per cent and swept through the island. Though not true in that case, the spread of an innovation often has nothing to do with the official research and extension system. Brammer reports many innovations originated by small farmers in Bangladesh and their rapid spread. 'The impression is gained of an unofficial research and extension network operating independently – even obliviously – of government programmes, often more practically oriented than the latter and, becuase of this, apparently more successful in terms of new adopters' (1980, p. 25). All too commonly, the unofficial research network is overlooked by all except small farmers themselves.

The best of both

In most countries of the third world, rural people's knowledge is an enormous and underutilised national resource. John Hatch has written that the small farmer's expertise represents 'the single largest knowledge resource not yet mobilised in the development enterprise', going on to say that "we simply cannot afford to ignore it any longer' (1976, p. 17). For Bangladesh, Hugh Brammer has observed that ignorance of the basic adaptive research carried out by farmers themselves implies waste of 'a tremendous resource of native talent and information which officials could use to amplify and accelerate their own research and development activities' (1980, p. 25). Brokensha, Warren and Werner consider that indigenous knowledge systems should be regarded as part of national resources, 'although so far nearly all nations have virtually ignored this national asset' (1980, p. 3). These statements are true in several domains besides agriculture with its dense research network of small farmers. In medicine, indigenous health practitioners are there already (Pillsbury, 1979), operating at the periphery but largely unconnected either with the modern medical system or with each other. In fishing, forestry, game, animal husbandry, and water resources there are also innumerable skilled and well-informed local experts.

The word 'expert' is used advisedly. Many activities undertaken by rural people and scientists are similar: they distinguish, name and classify entities in their environments: they observe, compare and analyse; they experiment; they attempt to predict. Enough has been presented to show, contrary to some professional prejudice, that there is much for outsiders to learn from rural people. The question now is to assess the relative strengths and weaknesses of outsiders' and rural people's knowledge, and to see how the strengths may be combined and the weaknesses neutralised.

Let us look at this from the point of view of their comparative advantages.

Some of the strengths and weaknesses of rural people's knowledge are embedded in their languages and concepts. What is perceived affects the language evolved to describe it; and language in turn provides concepts and categories which shape perception. Colour discriminations are an example. On the one hand, the colours actually seen by people vary: populations near the equator tend to have accumulations of pigment within the eye, acting as a protection against the potentially carcinogenic high incidence of ultra-violet and near-ultra-violet radiation, and this pigment attentuates short wave-length radiation. The lowered discrimination of shades of blues and greens is said to be reflected in the absence of terms differentiating these colours (Bornstein, 1975). On the other hand, the many fine distinctions of shades and patterns of brown among pastoral nomads which are used to describe and distinguish their animals are proverbial, and probably without equivalents in other languages. Colours illustrate nicely that different people can and do see and discriminate in different ways.

Other local words and concepts are inclusive rather than differentiating, combining categories which the outsider is trained to keep separate. These can be helpful. A simple example is a local term for a soil-vegetation association which an outsider would do well to adopt for its utility. Other words and concepts can lead to confusion. They may combine spatial, social and ecological dimensions in a broadly inclusive span. The Hausa word 'gàrii' means: 'human settlement or inhabited place, especially township or village-area (also country or chiefdom), community, inhabitants (of a settlement); (local) economy, including farmland, crops, weather and (local) sky' (Dalby, 1964). This and some similar words:

> ... are ecological concepts par excellence. If the words are hard to classify into the field of 'geography' or 'sociology', this is because their meaning applies at once to a place, to the social group which occupied the place, to that group's internal structure, to the relationship between the group and the place, and to the way in which the place has been moulded by the group
>
> (Langley, 1975, p. 97)

Appreciating this range of meanings is important if outsiders are to understand rural people's ways of thinking and to avoid misunderstanding. More practically useful will be words which discriminate finely or which describe stable associations which in other languages are kept separate.

Much of the relative strength of rural people's knowledge lies in what can be observed locally and over a sustained period, and in what touches directly their lives and livelihoods. Most obviously, this applies to their knowledge of customs and practices. Except where there has been systematic ethnography, this knowledge is superior to that available to outsiders. Descriptive and conceptual terms also provide points of departure for scientific investigation which may be more practical and useful than the externally determined categories of outsiders' knowledge. For example, a scientific soil survey in a semi-arid area can miss micro-environments with typical combinations of slope, soil, vegetation and micro-climate. These may be small places sheltered from the wind where the soil is deep enough for a few bananas to grow; or narrow strips, a few feet wide, of fertile alluvium, beside streams where vegetables can be irrigated; or the margins of seasonal ponds where crops can be planted. These may all have local names and be as critical for the livelihoods of local inhabitants as they are easy to miss for unobservant outsiders coming with their different preconceptions and categories.

Sometimes rural people's knowledge and that of outsiders are evenly balanced. In his study of local knowledge of the grasshopper, *Zonocerus*, Paul Richards found that community knowledge was equivalent to that of the scientific research team concerning what it ate, the degree of damage it did to cassava, and the tendency for eggs to be laid in certain sorts of places; but he found that community knowledge added to the knowledge of the scientific team concerning the dates, severity and geographical extent of some of the outbreaks, and the fact that the grasshopper was eaten and sold, and was especially important for women, children and poorer people (1980, p. 185). Rural people's knowledge is at its strongest with what is observable and its local *what*, *where*, and *when*.

Not surprisingly, then it is in agriculture that rural people's knowledge has its most marked local advantages, and that of outsiders has been at its weakest. The dangers of the Groundnut Scheme in Tanzania might have been foreseen more clearly if the local inhabitants had been carefully consulted about why they did not cultivate in the areas proposed for the project. The examples of inappropriate agricultural research on research stations, and harmful advice emanating from them, are legion – advice to grow crops on soils or in seasons which do not suit them; to apply more fertiliser than is profitable or justified by risk; to grow crops which threaten a family's food supply because of the land they take up or the labour they would pre-empt at a critical time; insistence on soil conservation measures which destroy fertility; advice to plant pure stands when mixed cropping makes more sense. It is also in agriculture that the strongest reversals have taken place, and where there has been most learning from rural people – through interviews, observing farmers' practices, surveys, on-farm trials, and on-farm experiments with farmers as colleagues. Professional outsiders' knowledge of agriculture has already gained much by trying to fit together what small farmers want and know and what formal scientific agricultural research can do.

Outsiders' scientific technology is superior in being able to measure precisely and examine microscopically, while rural people are usually relatively weak on measurement and unable to observe except with the naked eye. The precision of outsiders' statistics can, however, be gravely misleading. Rural people could probably have given much more accurate estimates of village-level grain losses in storage than those high figures which were so misleadingly believed for so long. (Rural people protect their grain in ways which scientists do well to study: in India, the berry of the neem tree is added by many people to their stored grain, and belatedly scientists have discovered the potential of an insecticide – azadirachtin – which can be extracted from the neem.) The weaknesses of measurement by rural people can also be exaggerated. Local standards and units of measurement exist and may incorporate important criteria which outsiders would otherwise miss. People can also be taught or helped to measure or quantify. Filipino farmers have been taught to map their fields, evidently without difficulty. There are ways in which farmers' estimates can be sharpened and more readily expressed, as with an adaptation of the West African mancala board by David Atteh and Paul Richards (Barker, 1979).

Despite these qualifications, the balance of advantage with precision and with the microscopic remains with outsiders' technology. Richards found in his *Zonocerus* investigation that community knowledge could not extend to findings concerning egg mortality under different conditions, or the possible role of chemical attractants in helping create and sustain egg-laying sites. These required precise quantitative data, experimental control, and sophisticated biochemical analysis (1979, p. 29). Nor is there any way in which rural people can identify a missing trace element in their soil. Again, in the health field, rural people have no direct way of knowing the aetiology of many diseases; they cannot observe bacteria, viruses, or even internal parasites. When it comes to malaria, the !kung San, so sophisticated in their plant lore and knowledge of animal behaviour, are reduced to the method of inference they so rightly distrust: for they believe that malaria is caused by a large caterpillar seen only in the rainy season when malaria is prevalent (Blurton Jones and Konner, 1976, p. 344).

Rural people's knowledge, in the form of beliefs and practices, is sometimes harmful according to the values of rural people themselves. It is true that there is much to be learnt from indigenous medical and psychiatric practices, and there is a large literature on ayurvedic, homeopathic and other local medicine, as well as on medicinal plants. But there are also some beliefs and practices in health and nutrition which make things worse, not better: reducing the fluid intake of children with diarrhoea, believing that less fluid going in will mean less coming out and so help to cure; the use of weaning foods which lack proteins and vitamins when cheap, nutritious alternatives are available; failure to recognise kwashiorkor because children with it can look robust and healthy (Fonaroff, 1975, p. 120); a belief that dehydration caused by diarrhoea is a separate condition, most commonly caused when a mother feeds her small infant after seeing a woman who has had a miscarriage (Lozoff *et al.*, 1975); not bringing children with diseases like chickenpox and measles for treatment for fear of making angry the goddess or her sisters who were believed responsible for the condition (Mather and John, 1973); the use of incisions or tourniquets to treat snake bites, thereby increasing morbidity (Warrell and Arnett, 1976, p. 331); or easily avoidable insanitary practices which spread disease.

It seems to be more in health and nutrition than in agriculture that harmful local beliefs and practices are found. It might be supposed that the incentive to observe, and to be effective, would be greater with what directly touches human welfare and survival in health and nutrition; but this does not seem to be the case. Several explanations can be advanced. In growing crops, there is a large population of plants from which to learn, and a few are expendable; but with people there are fewer and each one is precious. The causes of poor crop performance (drought, flooding, a pest, lack of soil moisture) or of good performance (good soils, manure, timely planting, weeding) are often only too obvious, compared with the invisibility of microscopic infections or the spread of disease. Again, learning from agricultural practices can also occur every season, whereas the care of a child through the stages of growth, even in large families, occurs less often, and sickness and malnutrition are spasmodic. Perhaps, also, sickness so engages the emotions that the experimental attitude is driven out. And the coping mechanisms for the awfulness of the illness and death of those who are close are social and spiritual, and so linked with social and spiritual rather than physical explanation.

But whatever the reasons, it is not sheer prejudice when outsider professionals see that rural people's beliefs and practices are sometimes harmfully wrong. Both outsiders' knowledge and the knowledge of rural people can be wrong. The key is to know which is wrong when. It would be as foolish here to do a complete reversal in favour of rural people's knowledge as it has been so often in the past to suppose that professional outsiders have a monopoly of insight.

There are many cases, especially in health and nutrition, where professional outsiders' knowledge can help rural people better to achieve what they want. Its strongest advantages lie in its capacity through the experimental method and through its command of resources and skills to generate new technology, and then to transfer it from one environment to another. Many examples potentially benefit rural people: inoculations against human and animal diseases; oral rehydration for small children with diarrhoeas; drought-resistant or drought-avoiding varieties of staple food crops; new methods for lift irrigation; new high-yielding varieties of crops; fertilisers and pesticides, and so on. There is a debate about who benefits, and how much. But the power of these techniques and artefacts of outsiders' science is beyond dispute.

Unfortunately, these dazzling capabilities blind outsiders. For originators and bearers of modern scientific knowledge, it requires a major effort to recognise that rural people's knowledge exists at all, let alone to see that it is often superior. The arrogance of ignorant educated outsiders is part of the problem. They do not know what rural people know and do not know that not knowing matters. Four positive examples can make the point.

First, with our friend *Zonocerus variegatus*, laboratory results showed that cassava leaves were by no means a favourite food of the grasshopper, yet it attacked them severely during the dry season. A local inhabitant gave a basic reason – that cassava is one of the very few plants with green leaves surviving in farm fields during the dry season. That may not seem such a remarkable insight but it might not have been available so quickly to the scientists on their own. Again:

> The most impressive overall understanding of the *Zonocerus* problem came from Kabba farmers, who explained its general incidence in recent years by reference to rainfall fluctuations, but in many cases accounted for its specific appearance on their farms as being due to colonization of neighbouring thickets by the herbaceous weed *Eupatorium odoratum*. These thickets do indeed appear to provide favourable breeding or feeding sites, and many Kabba farmers were anticipating advice which may emerge from studies on the biology and control of *Z. variegatus*, by cutting down *Eupatorium* and, in one or two cases, marking out and digging up egg-laying sites.
>
> (Barker *et al*., 1977, p. 46)

In this aspect, the farmers were ahead of the scientists in working on the same problem.

In a second example, a scientist, Peter Jones, was working on the bird pest *quelea*. He was travelling by Land Rover in Botswana with two !kung men:

Knowing of Jones' interest in quelea (he had been retained by the government to explore possible solutions to the serious quelea pest problem), the two men pointed out a low stand of thorn bushes which, at a distance, looked like any other but which, on close examination, proved to have been stripped of leaves on the distal few inches of their branches. The men said that this had been done by quelea, which were in the habit of preparing bushes in this way and then returning after a few days to rest on the ends of the branches. This observation, which was unknown to Jones, and which proved to be correct, enabled him subsequently to improve greatly the efficiency of his investigation and to collect at an early stage of the nesting cycle specimens previously inaccessible to him.

(Blurton Jones and Konner, 1976, p. 340)

A third case comes from Jeremy Swift's work with Wodaabe Fulani pastoralists in Niger. In order to benefit from their local knowledge, his research team asked the herders to draw maps, which the herders did without difficulty. The maps showed ecological units, as might be expected. But in addition, the Wodaabe mappers indicated several special zones. These were areas in which their cattle got night blindness in the dry season, and for this reason had to leave otherwise good pastures. They associated night-blindness with the absence of certain types of green plant, which fits the scientific explanation of vitamin A deficiency. It emerged that livestock service staff, who had been working in the area for 50 years, were not aware of this problem.[9]

Fourth, Hugh Brammer, whose intimate knowedge of Bangladesh agriculture has already provided material and insight for this chapter, can speak for himself:

Wheat cultivation on the Barind tract soils of Bogra District in the Northwest of Bangladesh provides an example of farmers jumping ahead of the scientists. Being a soil scientist, I had recommended that these impervious soils, puddled for transplanted rice cultivation in the monsoon season, were unsuitable for wheat cultivation in the following dry season. I found some farmers growing wheat on these soils. Not, of course, in the way in which wheat normally is grown, broadcast on the flat. These innovative farmers had made ridges by hand, as for cultivating potatoes – which are also grown in the area – and had sown two closely planted rows of wheat on each ridge. Also, they were irrigating the crop from dug wells or tanks (excavated ponds), applying frequent,

small amounts of water down the furrows so as to avoid waterlogging the soil. The crop growth appeared excellent, a view obviously shared by neighbouring farmers, because the practice has spread widely during the following two years. In retrospect, I recognised that cultivation on ridges and the application of frequent small amounts of water are the solution for cultivating dry-land crops on impervious soils, but I had not considered that the farmers would use horticultural techniques for cultivating a crop such as wheat.
(1980, p. 28)

These four instances have in common a feature of many scientific advances: the discovery of things which were not being looked for. The scientists working on *Zonocerus* can scarcely have been looking to the farmers to try out their control recommendations before they had even been formulated, yet that was what some were doing. Peter Jones travelling in Botswana did not know that his guides knew how *quelea* stripped leaves from the distal few inches of some bushes, preparing to return to rest on them a few days later. Jeremy Swift was trying to get the Wodaabe Fulani to map, not trying to find out about nightblindness in cattle. And Hugh Brammer may not have been looking for farmers growing wheat on soils he had recommended as unsuitable. But in each case, the unexpected was noticed and in each case the revelation was available from rural people. One may wonder how much goes unknown because of unseeing eyes, unhearing ears, professional conditioning, and the biases of rural development tourism. Neither rural people nor outsider scientists can know in advance what the others know. It is by talking, travelling, asking, listening, observing, and doing things together that they can most effectively learn from one another. For that, special attitudes and behaviour are called for from both parties but especially from the outsiders since it is with them that more of the initiative lies.

Finally, some of the greatest challenges are where both outsiders' and rural people's knowledge have been found wanting. Any list might include three great outstanding problems: the aetiology and prevention of diarrhoeas; sustained and stable small farming in marginal rainfed environments; and tragedies of the commons in the exploitation of natural resources. It is difficult to overstate the importance of these three. The diarrhoeas are major killers of children in the rural third world. Marginal rainfed environments, especially but not only in Africa, are supporting or failing to support more and more of the poorest and are undergoing what may often be irreversible degradation leading to deeper poverty. And tragedies of the commons are found in fisheries, forestry and pastoralism – through competition between rural people themselves and through appropriation by outside state and commercial interests: examples are the over-fishing and declining production of East African lakes, the destruction of India's forests, and the downward spiral of desertification and impoverishment, not only in the Sahel, affecting many of the world's 100 million pastoralists.

These three great outstanding problems are not alone in showing human knowledge, ingenuity and will still a long way from solutions. For such intractable issues, the joint use of professional outsiders' and rural people's knowledge, skills and resources may be the best way forward, combining the precise observations, measurements and experiments of modern science over a narrower and briefer range with the local knowledge and more extensive and continuous observations of rural informants and experimenters. The two types of knowledge complement each other; and together they may achieve advances which neither could alone.

For that to happen, power must shift. Knowledgeable rural people are disregarded, despised and demoralised by urban, commercial and professional values, interests and power. For them to be better able to participate, control and benefit requires reversals. Among these, one first step is for outsider professionals, the bearers of modern scientific knowledge, to step down off their pedestals, and sit down, listen and learn.

Notes

1 For expansion of these points, see Chapter 7, pp. 172–9.
2 This should not be taken as support for the sometimes quite silly attack on cash crops for small farmers. Most of the critics would, were they small farmers, be only too keen to grow cash crops. My criticism is against the balance of research and extension, which has neglected the subsistence side.
3 See McDowell and Hildebrand, 1980, especially pp. 57–62 on 'Barriers to Integrating Livestock in Farm Systems Research', for parts of this paragraph and for other relevant points.
4 But see French, 1970; McDowell and Bove, 1977; and Sands and McDowell, 1979.
5 See Chapter 7, pp. 175–6.
6 Much relevant analysis and much empirical data, are to be found in Brokensha, Warren and Werner's book *Indigenous Knowledge Systems and Development* (1980), on which I draw extensively in this chapter.

7   As far as I know, Noble Savages were always thought of as men.
8   Personal communication. Such patterns have also been established for temperate zones. See *Science*, 137, 1962 for articles by Bradley, Woodbury and Brier for the USA (p. 178) and by Adderley and Bowen for New Zealand (p. 749).
9   The subsequent experience deserves a footnote. The vitamin A treatment is easy. The research team obtained some vitamin A and took it to a Wodaabe camp where cattle had nightblindness, and approached a cattle-owner. He was willing for his cattle to be treated, but asked for only half of them to be treated so that he could observe the effects and compare them with the untreated half.

CHAPTER FIVE
Integrated rural poverty

We have no power to talk in front of the rich, like the Chairman. We are afraid of them. We are always looked down upon and scolded. So we never know what they are writing and doing.

> A landless labourer in Bangladesh (BRAC, 1979, p. 20)

We used to go to people to hire us for the brewing of beer and for collecting some water but now they are refusing to help us. There is nowhere we can go for help. If you have nothing, you have nothing and it ends there.

> The eldest daughter in a poor household in Botswana (Henderson, 1980, p. 226)

Sometimes you are overcome by weeds through illness or accidents.

> A Gambian villager to Margaret Haswell (1975, p. 44)

I do not wish to speak to you about these things, for my situation is so miserable and I am so desperate that I cannot go on talking of them. It is not words that can change my life, but a change in my country...

> Interview in Nepal, reported in Blaikie, Cameron and Seddon, 1979, p. 48

Outsiders' comfortable views of the poor as improvident, lazy, fatalistic, ignorant, stupid and responsible for their poverty, are reassuring but wrong. Case studies show that poor rural people are usually tough, hard-working, ingenious and resilient. They have to be to struggle against five interlocking disadvantages which trap them in deprivation: *poverty* itself, *physical weakness*, *isolation*, *vulnerability*, and *powerlessness*. All are important, but vulnerability and powerlessness especially deserve more recognition and analysis.

Vulnerability reflects lack of buffers against contingencies such as social conventions (dowry, bridewealth, weddings and funerals), disasters, physical incapacity (sickness, the child-bearing sequence, and accidents), unproductive expenditure, and exploitation. Contingencies often force poverty ratchets, entailing the irreversible loss or sale of assets, making people poorer and more vulnerable to becoming poorer still.

Powerlessness is reflected in the ease with which rural elites act as a net to intercept benefits intended for the poor, in the way the poor are robbed and cheated, and in the inability of poorer people to bargain, especially women, and those who are physically weak, disabled or destitute. Altruism and generosity are also found, but reciprocal relations and traditional supports for the poor are rarer and weaker than in the past. There are environments where greater prosperity has improved the material conditions of life for all except the most indigent and unfortunate, but there remain hundreds of millions of people for whom the trends are in the other direction, moving down into deeper and more tightly integrated poverty.

Outsiders' views of the poor

Outsiders' views of the poor are distorted in many ways. Lack of contact or communication permits them to form those views without the inconvenience of knowledge, let alone personal exposure. Poor people are rarely met; when they are met, they often do not speak; when they do speak, they are often cautious and deferential; and what they say is often either not listened to, or brushed aside, or interpreted in a bad light.[1] Any attempt to understand the poor, and to learn from them, has to begin with introspection by the outsiders themselves. We have first to examine ourselves and identify and offset our preconceptions, prejudices and rationalisations. Above all, we have to treat with suspicion beliefs and interpretations which we find comforting, and which purport to justify our relative affluence and the relative poverty of others.

The most reassuring view is that the poverty of others is part of a divine order. This idea is embedded in popular Hinduism and the belief that position in the caste hierarchy is determined by the law of karma, according to which the advantages and hardships of this life are a consequence of the degree of merit of past actions in a previous life. But it is not only Hinduism which justifies, or has justified, social inequality, the coexistence of rich and poor. In the much quoted words of Mrs Alexander's Victorian Christian hymn:

> The rich man in his castle,
> The poor man at his gate,
> God made them, high or lowly,
> And order'd their estate.[2]

The American ideology of success, dominant until the Great Depression of the 1930s, was another convenient belief for the better off: it regarded wealth as a reward for Puritan virtues such as honesty, industry, sobriety, self-discipline, neatness, cleanliness and punctuality, and saw poverty as the converse. Nor are these beliefs dead today. To the contrary, in Britain at least, the idea that the poor are to blame for their poverty has been widespread. A survey in 1976 in 9 countries of the EEC, including Britain found that 27 per cent of respondents in Britain, compared with only 14 per cent for the EEC as a whole, were 'poverty cynics' (CEC, 1977, p. 88)[3], that is, they were defined as people who

> ... rarely or never see poverty around them. When they mention it, they imply culpability – if poor people exist, it is because they are lazy or lack will-power and they or their children could well escape from this situation. As far as the cynics are concerned, there is no great need to reduce social inequality and the authorities are doing quite enough – if not too much.
>
> (*Ibid.*, p. 80)

Such beliefs are common in many cultures. In some cases they have antecedents in the racial ideologies of colonialism, and in the colonial view of the native as improvident, lazy and fatalistic. They are to be found among long established elites like those of Bombay or Buenos Aires. A parallel can be found between the old American myth of the opportunity of an open frontier and the new African myth of the opportunities of education: those who have made it to good urban jobs owe their success to diligence rather than influence, and the remaining rural poor are those who did not work hard enough. The greater the inequalities and cognitive distance between urban, educated haves and rural, less educated have-nots, so the commoner may be beliefs which hold the rural poor responsible for their poverty. Such beliefs so happily rationalise the haves having and the have-nots not having, that it would be odd if it were not so.

Now outsiders cannot help bringing with them whatever ideological baggage they have, and it is difficult to avoid choosing and collecting evidence that fits into it. What they can do is travel lightly, asking open-ended questions, listening, observing, revising their ideas, and above all doubting and criticising themselves. Passionate moral indignation, however necessary as a driving force for action, can be an impediment, just as can cold conservatism. Evidence will always be selected, but to struggle closer to the truth a certain dispassion helps.

The evidence itself is imbalanced. It has been generated by top-down, centre-outwards processes of learning. The rural poor are scanned in misleading surveys, smoothed out in statistical averages, and moulded into stereotypes. This scarcely helps an outsider, who starts by being affluent and urban, to make the long leap of imagination and see and feel the world from within the skin of a poor rural person. Nor has social science research helped as much as it might. Studies count surface phenomena, or conceptualise; not many reveal the world view, the problems and the strategies of particular poor individuals and families. Yet for generalisation, one needs to start with the raw material, with cases, with people.[4] Those which I have examined reflect many differences of culture, ecology and social, economic and political relations. They also reveal common features.

The evidence does not support the view of poor rural people as improvident, lazy and fatalistic. What does emerge is that some do sometimes behave in ways which can be thus interpreted. They may not save, may not always be visibly working, and may appear to accept fate passively. But there is evidence for interpretations of this behaviour other than moral defects in character. 'Improvidence' – the failure to save and invest – can reflect pressing needs for immediate consumption, a backlog of essentials needed, insecurity of land tenure, and the likelihood that any saving would attract the attention of begging relatives and social predators. 'Laziness' conserves energy: those who live near the margin hoard their strength and ration their effort. ('I pick fennel when it's in season, and we eat it at home. When there's no fennel and nothing to do, I go to bed' (Dolci, 1956, p. 261).) 'Fatalism', too, can be seen as an adaptation: like resting, it conserves physical and mental energy. It is also prudent. Appearances of powerlessness, unawareness, and acquiescence may be a condition for survival, for a chance of casual work, for the next loan from the trader, landlord or money-lender, for freedom from petty persecution and expropriation.

Nor does the evidence support the belief that the rural poor are ignorant and stupid. The depth and validity of rural people's indigenous technical knowledge is one dimension. Another is the understanding poor people have of why they are poor (Freire,

1968; BRAC, 1979, 1980; Malik, n.d.). This is clearer and more detailed than some outsiders might suppose. Apparent ignorance and stupidity are part of the strategy of lying low. Indian tribals asked by Baljit Malik why they kept being polite to officials who visited them, always agreeing to everything, replied with the saying: 'If the circumstances so demand, keep saying YES; if someone asks whether you saw a cat carrying a camel in its mouth, say YES!' (Malik, n.d., p. 13). Questionnaire surveys badly administered can also generate spurious figures which grossly understate the knowledge of rural people (see pp. 55–6). Ignorant and stupid poor people are often the creation of ignorant and stupid outsiders.

In correcting the prejudiced view of poor rural people as culpably improvident, lazy, fatalistic, ignorant and stupid, the pendulum can swing too far. The poor, the landless, the illiterate and the oppressed can be idealised. A modish account of peasant decision-making could read as though the subject were a sublime incarnation of Rational Man. There are some stupid and lazy poor people, just as there are stupid and lazy rich, and poor people can miscalculate, make mistakes, get drunk, and forget things, as others do. But the evidence speaks for itself. Again and again and again, observers have remarked on the toughness, application and ingenuity of the poor. Leela Gulati's (1981) case studies of five poor women in Kerala describe in detail the gruelling physical work and long hours they undergo, and that on astonishingly low calorie intakes. One of Dolci's informants said 'I cudgel my brains day after day wondering what to do. To get by, you've got to scrape a bit here, a bit there. If you don't you die' (Dolci, 1966, p. 267). Rural squatters in Kenya have a resilience often born out of desperation (Mbithi and Barnes, 1975, p. 165). John Hatch was in rural Peru when small farmers were hit by a flood disaster:

> I briefly toyed with the idea of documenting the disaster itself. But then I was captured by an even more impressive phenomenon: the relentless tenacity of the small farmers themselves; their amazing ability – without disaster assistance from any source – to bounce back and begin land preparation for a new crop. Like pawns on a chessboard, their only option was to move forward.
> (1976, pp. 14–15)

People so close to the edge cannot afford laziness or stupidity. They have to work, and work hard, whenever and however they can. Many of the lazy and stupid poor are dead.

To get beyond stereotypes and counter-stereotypes requires comparative analysis of the micro-detail of rural poverty. This is not easy. With some evidence, there are problems of bias in what poor people say, and in cases selected for writing up. Case histories of people and families have also been few: the traditions of research and scholarship have absurdly neglected and undervalued the particular, the non-statistical, and the easily obtained. Social anthropologists have contributed, but it has been an economist, Leela Gulati, who has most decisively shown what can be done with painstaking research to record the lives of poor rural people. Her study (1981) of five poor rural women in Kerala breaks new ground in its detail and persuasiveness and is a model of the sort of investigation that is needed.

But even with good cases as raw material for analysis, there are still problems of interpretation and ideology. No analysis is 'objective'. Eclectic pluralism is itself an ideology. I can only say that I have tried to review case study and other evidence and to see what categories and generalisations they generate. The outcome, which follows, suggests that there are striking common dimensions in conditions, activities, relationships and strategies of the poorer rural people in different regions and countries.

Clusters of disadvantage

A description of the condition of poor rural people might start with communities or with individuals. Starting with communities would have the advantage of distinguishing two types of situation: those where the poverty of whole communities is linked to their remoteness or inadequate resources or both; and those where there are marked differences of wealth and poverty within the same community.[5] Starting with individuals would have the advantage of pointing to the disadvantages of females in many societies, sometimes from the moment of birth. These two dimensions – of location and resource base, and of gender – are significant, and qualify all that follows: some communities are much poorer than others, and more uniformly poor; and women are usually, but not always, poorer than men.

It is, however, households that are the common, and increasingly distinct economic entities for production, for earning, and for sharing consumption. The approach here is to try and identify clusters of disadvantage of households, separate them, and then see whether, and if so how, they are connected. This could be done in many ways, and no particular merit is claimed for the categories which follow. Readers can list their own. But it is useful to dissect evidence and not to allow the term 'poverty' to cover all aspects of disadvantage, but only those – lack of wealth or assets, and lack of flows of food and cash – to which it properly refers. To make a start, five clusters of disadvantage can be described – poverty, physical weakness, vulnerability, isolation, and powerlessness. These can be presented as a composite sketch of the household.

i) *The household is poor.* It has few assets. Its hut, house or shelter is small, made of wood, bamboo, mud, grass, reeds, palm fronds or hides, and has little furniture: mats or hides for sleeping, perhaps a bed, cooking pots, a few tools. There is no toilet, or an insanitary one. The household has no land, or has land which does not assure or barely assures subsistence or which is rented or sharecropped. It has no livestock, or has only small stock (hens, ducks, goats, a pig . . .) or a few weak cattle or buffalo. The household borrows from neighbours, kin and traders, and is in short-term or long-term debt. Clothes are few and worn until they are very old. Family labour has low productivity: if it farms, its land is marginal or small; if it does not farm, it has little or no control over the means of production, and its main, often only, productive asset is the labour of its members.

The household's stocks and flows of food and cash are low, unreliable, seasonal and inadequate. The household is either locked into dependence on one patron, for whom most work is done, or contrives a livelihood with a range of activities which reflect tenacious ingenuity in the face of narrow margins for survival. Food or cash obtained meet immediate needs and are soon used up. All family members work when they can, except the very young, the very old, the disabled, and those who are seriously sick. Women work long hours both at domestic tasks and outside the home. Returns to the family's labour are low, and in the slack seasons often very low, if indeed there is any work then at all.

ii) *The household is physically weak.* There is a high ratio of dependents to able-bodied adults. The dependents may be young children, old people, the sick, or handicapped. The ratio of dependents to able-bodied adults is high for one of several reasons: because there is no man and the household head is a woman with responsibilities for child care, food processing, cooking, drawing water, collecting firewood, marketing and domestic chores, besides earning a livelihood for the family; or because of the stage of the domestic cycle when there are small children demanding time, food and care but not yet contributing economically; or because adults have been permanently weakened or disabled by accident or illness; or because of early deaths of other adults; or because active adults have dispersed or migrated to escape poverty or debts or to survive. The adults are seasonally or continuously pressed for time and energy. The household is seasonally hungry and thin, and its members weakened by interactions of parasites, sickness and malnutrition. Pregnancy, birth and death are common. Birth weights are low. All have small bodies, stunted compared with their genetic potential.

iii) *The household is isolated.* The household is isolated from the outside world. Its location is peripheral, either in an area remote from town and communications, or removed within the village from the centres of trading, discussion and information. Often illiterate and without a radio, its members are not well informed about events beyond the neighbourhood. Its children do not go to school, or go and drop out early. Its members either do not go to public meetings, or go and do not speak. They do not receive advice from extension workers in agriculture or health. They travel only to seek work or to beg from relatives. They are tied to their neighbourhood by obligations to patrons, by debts, by immediate needs that must be satisfied, or by lack of means for travel.

iv) *The household is vulnerable.* The household has few buffers against contingencies. Small needs are met by drawing on slender reserves of cash, by reduced consumption, by barter, or by loans from friends, relatives and traders. Disasters and social demands – crop failure, famine, a hut burning down, an accident, sickness, a funeral, a dowry, brideprice, wedding expenses, costs of litigation or of a fine – have to be met by becoming poorer. This often means selling or mortgaging assets – land, livestock, trees, cooking pots, tools and equipment, ration books, jewellery, a standing crop, or future labour, often on distress sale or usurious terms. Vulnerability is heightened during wet seasons when food shortages, sickness and agricultural work coincide, and is acute when rains and agricultural seasons fail. The family is especially prone to sickness and death.

v) *The household is powerless.* Ignorant of the law, without legal advice, competing for employment and services with others in a similar condition, the household is an easy victim of predation by the powerful. It has inherited or descended to low social status. Its position is weak in negotiating terms for the use of its labour or the sale of its produce or assets. It is easily exploited by moneylenders, merchants, landlords, petty officials and police. Aware of the power of the richer rural and urban people and of their alliances, the household avoids political activity which might endanger future employment, tenancy, loans, favours or protection. It knows that in the short term accepting powerlessness pays.

To some, this sketch may appear exaggerated. There are exceptions. Some poor families are less weak physically than described. Poor people in pastoral populations have different patterns of deprivation. Some of the strongest qualifications apply in East Africa where vertical patron-client relations, chronic indebtedness, and exploitation of the poorer peasants by those who are less poor, are much less in evidence than say, in Bangladesh. Other countries spring to mind where only a minority of the rural population might fit this description: perhaps Taiwan, Korea, China. Nor is isolation always a trait. In Sri Lanka there is near-universal primary education. In Kerala, many rural people read and discuss newspapers daily. And accepting Goran Hyden's (1980) analysis for Tanzania, small can be powerful: small peasants, seeking to retain their independence of bureaucracy and the manifestations of the state, adopt isolation as a strategy, avoiding becoming powerless by avoiding certain types of contact.

Such qualifications do not invalidate the general description. Rather, they identify places and conditions in which some of the forms of disadvantage have been avoided, overcome, or made use of. Any immediate urge to reject the description might be tempered by reflection on the anti-poverty biases (pp. 13–23). Most poverty, quite simply, goes unseen; and where perceived, is only seen in one or a few dimensions. My best judgement is that for the great concentrations of rural poverty in South and Southeast Asia, in Africa, and in Latin America, most of the description holds true, applying broadly to perhaps a half to three quarters of the rural people in the third world.

The deprivation trap

Still examining poor households and their immediate environments we can see that these clusters of disadvantage interlock. This is variously described as the vicious circle of poverty, the syndrome of poverty and the poverty trap. We can go further than saying people are poor because they are poor because they are poor. Linking the five clusters (Figure 5.1) gives twenty possible causal relations, which in their negative forms interlock like a

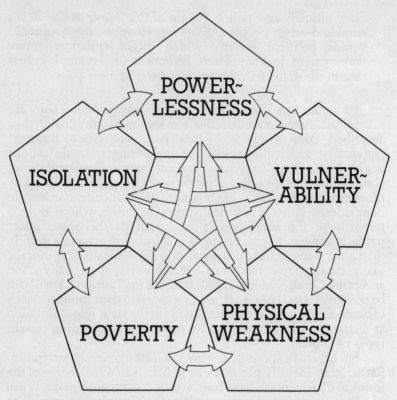

Figure 5.1 The deprivation trap

web to trap people in their deprivation. The strength of these linkages varies, but they can be illustrated by starting with each cluster in turn.

Poverty is a strong determinant of the others. Poverty contributes to physical weakness through lack of food, small bodies, malnutrition leading to low immune response to infections, and inability to reach or pay for health services; to isolation because of the inability to pay the cost of schooling, to buy a radio or a bicycle, to afford to travel to look for work, or to live near the village centre or a main road; to vulnerability through lack of assets to pay large expenses or to meet contingencies; and to powerlessness because lack of wealth goes with low status: the poor have no voice.

The *physical weakness* of a household contributes to poverty in several ways: through the low productivity of weak labour;

through an inability to cultivate larger areas, or to work longer hours; through lower wages paid to women and to those who are weak; and through the withdrawal or weakening of labour through sickness. It sustains isolation because of lack of time or energy to attend meetings or to seek information, especially for women because children make travel difficult. It accentuates vulnerability by limiting the ability to overcome a crisis through harder work, new activities, or negotiations for help. It contributes to powerlessness through the lack of time or energy for protest, organisation, or political activities: sick and hungry people dare not bargain hard.

Isolation (lack of education, remoteness, being out of contact) sustains poverty: services do not reach those who are remote; illiterates cannot read information of economic value, and find it difficult to obtain loans. Isolation goes with physical weakness: remote households may have a high level of migration of the able-bodied to towns or to other rural areas. Isolation also accentuates vulnerability – remote marginal areas are more liable to crop failures, and are less well provided with services to handle contingencies like famine or sickness; illiterates also find it harder to register or acquire land and are more easily cheated of it. And isolation means lack of contact with political leaders or with legal advice, and not knowing what the powerful are doing.

Vulnerability is part of many of the links. It relates to poverty through the sale or mortgage of productive assets; to physical weakness because to handle contingencies, time and energy have to be substituted for money; to isolation through withdrawal – whether spatial (to a more distant marginal area) or social (to fewer reciprocal relationships) – following shocks and contingencies; and to powerlessness through the dependence on patrons to which it gives rise.

Finally, *powerlessness* contributes to poverty in many ways, not least through exploitation by the powerful. It limits or prevents access to resources from the state, legal redress for abuses, and ability to dispute wage or interest rates; and it entails weakness in negotiating the terms of distress sales, and only feeble influence on government to provide services for the poorer people and places. It reinforces physical weakness, because time and energy have to be devoted to queuing for access, because labour obligations to patrons reduce labour available for household production or other earning; and because relief food supplies in time of famine may never be obtained because people are powerless to demand what is meant for them. Isolation is linked with powerlessness through the inability of those who are powerless to attract government aid, schools, good staff, or other resources. Powerlessness also makes the poor more vulnerable – to sudden demands for the repayment of loans, to threat of prosecution and fine or imprisonment, or to demands for a bribe in a dispute.

The five clusters and their linkages could be much more fully explored. They could be further illustrated, tested and modified against cases. But two of them are relatively well accepted and understood: poverty; and isolation, both spatial and informational. A third – physical weakness – is subject to an important and fascinating debate on which it would be premature to pass judgement. This suggests, in brief, that estimates of minimum calorie requirements have been high, and that many people who would be classified as malnourished in terms of height for age ('stunting'), are normal in terms of weight for height, and may in principle be described as 'small but healthy' (Seckler, 1980a; 1980b, p. 225). Much more needs to be known about all this.

No doubt the next decade will provide both intellectual excitement and useful understanding about human nutrition, health, growth, mental and physical capacities, and qualities of life and experience. In the meantime, any scaling down of estimates of the numbers of people who are 'malnourished' in the world must be sharply qualified by a recognition of widespread seasonal stress of shortages and starvation on families which are at risk. In tropical climates this stress is common during the rains and before the first crops are harvested, when food shortages, hard work, poor child care, and high disease incidence interact, and when urban-based professionals are least likely to have contact with the poorer rural people (Schofield, 1974; Longhurst and Payne, 1979; Chambers, Longhurst and Pacey, 1981; Chambers, 1982). Minimum calorie requirements may have been somewhat exaggerated, but seasonal stress is still seriously underperceived by outsiders.

Of the other two clusters, vulnerability has been curiously neglected, and powerlessness is a key problem which many outsiders find it uncomfortable to face. I shall therefore examine these two in more detail.

Vulnerability and poverty ratchets

Households become poorer by loss of assets. To meet small needs, ready cash or barter may be used, or small loans from neighbours, kin, patrons or traders. But to meet big needs, or for small needs at times of seasonal shortage or general crisis, it is often necessary for poor people to mortgage or sell capital assets. Where these events entail debts at high interest rates or loss of productive wealth, they can have ratchet effects, like movements down past a cog[6] which are difficult or impossible to reverse, making poor people permanently poorer. The poverty ratchet – the loss of assets or rights which it is difficult to reverse – may be forced by a slow build-up of pressures which pass a threshold, by an expenditure which is foreseeable but large, or by a sudden crisis. Contingencies which force poverty ratchets are of five main types: social conventions; disasters; physical incapacity; unproductive expenditure; and exploitation. These will be described separately, but their force is greatest when they combine, either at the same time or in sequence.

Social conventions which make heavy demands include dowry, bridewealth, weddings and funerals. In the Indian subcontinent these were a favoured explanation of indebtedness during the British raj (e.g. Darling, 1947, pp. 48–59). Possibly some of this emphasis should be discounted as it fitted an exaggerated stereotype of the improvident and extravagant peasant; but there can be no doubt that the costs of ceremonies and social transactions did often drive people deep into debt. Nor may the burden of such demands always have eased. In two villages in South India Scarlett Epstein records a rapid change from bridewealth to dowry as part of the process of Sanskritisation whereby lower castes adopt the practices and attributes of higher castes. For example:

> Boma, one of the poorer Peasants, who used to work in the Wangala factory plantation, told me that the recent marriage of his eldest daughter had landed him deeply in debt. He had tried very hard to find her a husband without having to give dowry, but without success. The girl was already 15 years old and had reached puberty. His wife was urging him to arrange a marriage. He finally settled the wedding by paying Rs 1 000 dowry to the groom's father, giving a watch and clothes to the groom worth Rs 500, buying clothes and jewellery for Rs 850 for his daughter and spending Rs 1 200 on the actual wedding ceremony and accompanying feast – about Rs 3 550 altogether. He has only two acres of wet land and therefore hardly produces enough to meet his current household needs let alone to pay such heavy marriage expenses. He has three more younger daughters and dreads the time when their turn comes to get married.
> (Epstein, 1973, p. 196)

Dowry makes the father of many daughters an object of pity in

115

societies where it is customary. Small farmers may have to sell land to raise it. Not surprisingly, some of the poorer peasants in one of Epstein's villages were arguing against dowry on the grounds that the bride's family was losing an important productive helper while the groom was gaining one in his new wife. Indeed, of these social expenditures, bridewealth is the least impoverishing, and can be seen in one sense as an investment for the groom and his family.

In much of Africa, expenditures on bridewealth, marriages and funerals make similarly heavy demands on resources. Polly Hill, recording remarks volunteered in Northern Nigeria on 'other people's poverty', includes 'He sold all his farmland to meet his marriage expenses' (1972, p. 148). For the Giriama of Coastal Kenya, David Parkin lists bridewealth and funerary expenditure among important contingent demands, and notes that their costs had risen greatly. 'More than being simply obligatory, the funerals must be lavish occasions, the magnificence of which match, and frequently exceed, the resources of the nominal sponsors' (1972, p. 59).

Disasters take many forms. They may be directly man-made: theft of livestock, tools or jewellery; the burning of a hut; and war and persecution which instantly impoverish by destruction or seizure of crops and animals and by driving peasants off their land and away from their rights in resources. A household can be hit by the death of a cow, buffalo, mule, pig, goat[7], or other animal. Other disasters take a natural and widespread form: floods; droughts; epidemics of plant, insect and animal diseases; and famines. Statistical estimates for the reasons for disposal of assets are rare[8], but famine may be the most common trigger for sales of land and livestock. Where many people have some land but nothing to eat, as in Bangladesh in 1974, sales of land become epidemic as they surrender property on almost any terms in order to obtain food and survive. Similarly, for people with cattle, as in Dodoma in Tanzania in 1969 (O'Keefe and Wisner 1975, p. 36), a sudden rise in sales can reflect the passing of a threshold of lack of purchasing power beyond which loss of capital assets is accepted as the necessary high cost for survival.

Physical incapacity takes three forms: sickness; the child-bearing sequence of pregnancy, childbirth, and the post-natal period; and accidents. Pregnancy and some sicknesses build up gradually but much sickness and almost all bodily accidents have a sudden impact.

The effects of physical incapacity are two-fold. First, the weakening or loss of labour and earning power of an adult reduces or stops the flow of income and food into the household.

Single-adult female-headed households are most vulnerable, but in larger households too, effects can be severe especially if more than one member is incapacitated. Second, treatment almost always has to be paid for, whether for sickness, birth attendance, or injuries. With both traditional and allopathic medicine the sums vary widely, but they are often large. With traditional practitioners, loans are often possible and there may be an option of repayment by labour or in kind. With allopathic medicine, immediate hard cash is more likely to be needed for transport, medicines, hospital care, bribes, and food and sustenance both for the sick person and an accompanying family member. Offerings to gods, priests, shrines or temples may also be made. If the sickness coincides, as many do, with the tropical wet season (diarrhoeas, malaria, dengue fever, guinea worm disease, skin infections, snake bite), the costs of production and earning foregone from agriculture may be high and will be reflected in subsequent shortages of food and cash in the hungry period before the next harvest.

A prolonged illness can impoverish utterly. Interviews with working women, among the poorest of the poor in rural Bangladesh, suggest that one sequence starts with the sickness of a husband. As the sickness continues, assets are sold to pay for treatment, bit by bit, down even to the last small rings and jewellery. In the end, the husband dies, and his widow and children are left destitute. Similarly an accident – a back injury, a broken limb, a pulled muscle, a damaged hand or foot – can so weaken the labour and earning of a household that it sets the household on an irreversible downward slide.

Unproductive expenditure takes many forms. It can involve drink, drugs or other expensive or debilitating consumption. It can combine bad judgement and bad luck in various mixes, as with failures in business, litigation and gambling. It can take the form of bribes or inducements which are inadequate or which do not pay off. It can be made for apprenticeship or training which is not completed or otherwise does not lead to benefits. A common pattern is that capital is lacking to make productive what has already been invested – to buy stock for the small shop, to pay for the licence renewal, to sink the well deeper to reach water, to buy pesticides to protect the crop to which irrigation, labour and fertiliser have already been applied. Or again, the expenditure can be on an asset which is not directly productive – a better house, jewellery, a radio, shoes and clothes. A major cost for many households is education for their children even when it is nominally free: outlays are often quite heavy – for textbooks, uniforms, the sports fund, the building fund, and presents for the teacher; and the withdrawal of child labour also has costs. If education does not pay off as an investment, it too can make people poorer. And whatever the unproductive expenditure, if money is borrowed to finance it, interest payments and other obligations deepen the improverishment.

Exploitation includes excessive demands and illegitimate acts by the powerful. It has many forms. The exorbitant interest rates of money-lenders are sometimes explained in terms of high risks of non-repayment, but the astonishingly high rates of between 100 and 200 per cent or more per annum which are found in many parts of the world cannot be justified when so many of the debtors are trapped in their physical and social environment and unable or unlikely to run away. The use of trickery or force to cheat people of land, livestock, produce or access to communal resources is another form of exploitation. Again, poor people can have to find large sums to buy off police prosecution, to get out of illegal custody, to secure a hearing in a case, or to obtain a resource from the public sector. Intimidation, blackmail, and violence are overt means of exploitation, but much more widespread are its many less explicit, more subtle forms in which poor people accept bad deals for fear of loss of favour or reprisal.

To carry the discussion further, let us now examine some examples of poverty ratchets from different countries and regions. Those that follow are not representative in any statistical sense. I have no reason to doubt the accuracy of any of them, but it is as well to bear in mind that it is the more dramatic and disastrous events which are remembered and recounted. Although accounts of personal misfortune rarely lose in the telling, these all have about them the ring of truth.

In a highland village of Mexico, the Martinez family was continuously in debt and preoccupied with making ends meet. For Pedro, the father, as for most of the villagers, getting enough money for food and clothing from one harvest to another was the all-absorbing, never-solved problem (Lewis, 1959, p. 40). '...the really bad times were when there was a serious illness in the family. Then they had to sell nearly everything, sometimes all their young turkeys or a grinding stone, sometimes a mule' (*ibid.*, p. 43). On the occasion described, Pedro had recently borrowed money for surgery in hospital. Now he had just sold a mule to pay off a debt:

> ...and it infuriated him to think that he had to sell it for only 300 pesos when it was easily worth 450. And now he had only one mule left. This meant that the boys could bring only half the usual amount of wood down the mountains and that

there would be little left to sell after Esperanza took what she needed. Besides, during the plum season the boys could earn only half of what they had the year before hauling crates of fruit to the railway station. And at harvest twice as many trips would have to be made to bring the corn down from the fields.

(*Ibid.*, p. 40)

In the Philippines, the landless Sumagaysay family had a similar experience. The family head is Tiyo Oyo and his wife Tiya Teria. Antonio Ledesma records a pressing emergency in 1969:

Tiyo Oyo was striken by El Tor (a mild form of cholera) for a month. He had to be brought to the hospital in Pototan. The week's stay in hospital cost the family P120, with food not yet included. Another P130 had to be provided to buy dextrose when Tiyo Oyo was in a critical condition. Fortunately, one of the drugstores in Pototan agreed to provide a guarantee for the Sumagaysays in the hospital. To cover the expenses, Tiya Teria had to sell their carabao (buffalo) for P330 to another small farmer in the barrio. The carabao was already in full working condition, and under normal circumstances could have been sold for more than twice the amount received by the Sumagaysays. Moreover with the carabao, Tiyo Oyo would still have been able to plow other farm parcels for P10 a day instead of working as a pure manual labourer for the current wage rate of P6 a day . . . . In that sense, parting with the carabao meant parting with their last capital investment in farming. Buying a new carabao today would be unthinkable with the current market value of a working carabao estimated by barrio people themselves at P1 000–1 500.

(1977, p. 27)

The uncontrollable lot of the South Italian peasants has been described by E. C. Banfield (1958, quoted by Seligman, 1975):

What for others are misfortunes are for him calamities. When their hog strangled on its tether, a labourer and his wife were desolate. The woman tore her hair and beat her head against a wall while the husband sat mute and stricken in a corner. The loss of the hog meant they would have no meat that winter, no grease to spread on bread, nothing to sell for cash to pay taxes, and no possibility of acquiring a pig the next spring. Such blows may fall at any time. Fields may be washed away in a flood. Hail may beat down the wheat. Illness may strike. To be a peasant is to stand helpless before thesepossibilities.

In rural Bangladesh, an accident led to a poverty ratchet of a different kind. According to this account, the son of an informant broke his leg and the father took his son to the Tangail hospital for treatment. The doctors there said they had no supplies or instruments but that if the father could pay them 250 Taka, they would arrange for treatment. The father said that by selling his belongings he might be able to give them 50 Taka.

> The doctors told him this was too small a sum and that they could do nothing for that amount. Angry and disgusted, the man took his injured son back to the village where he was treated with whatever traditional medical knowledge they possessed ... the boy is now lame and will be a cripple for the rest of his life ... [and] now ... cannot lead a productive life. He will not be able to work in the fields but still, he will have to eat.
>
> (Jansen, 1978, pp. 27–28)

Again in Bangladesh, a poor householder inherited 0.19 acres from his father. On this land he grew cane and reeds from which he and his wife wove mats for sale. But during the 1974 famine rice went from 3 taka a seer to 10 taka a seer and he felt the only way he could raise enough money to feed his family was to sell his land. Now, having to buy the raw materials he formerly grew himself, the economics of mat-making are harder for him and his family (Jansen, 1978, pp. 19–20).

A final, more general example, comes from famines in West Africa. Polly Hill writes of a village in Northern Nigeria that:

> The people of Batagarawa have vivid memories of four famines: Malali (1914), Kwana or Kona (1927), 'Yar Balange (1942) and Uwar Sani (1954) – the dates all relating to the pre-harvest months of the year following the crop failure ... many migrated following Uwar Sani (1954) – it being 'after suffering that you migrate'. Those whose grain stocks were exhausted were obliged to sell their farms to others more fortunate – farm prices fell very low during 'Yar Balange (1942) and Uwar Sani (1954).
>
> (1972, p. 231)

These six examples – from Mexico, Italy, Bangladesh, the Philippines, and Nigeria – present some types of contingencies, and some evidence for dissection. But they are weak both cross-sectionally and longitudinally. They do not tell us about the relative importance in a community of different impoverishing events; nor do they show, except a little with Don Pedro, how sequences of events may affect a particular household. Let us examine these dimensions in turn.

Five studies have investigated the different reasons given for selling capital assets or taking debts. In chronological order of fieldwork these are by F. G. Bailey for sales of land in Bisipara village in Orissa in India; by David Parkin for sales of land and palms in Tsakani in Kilifi District in Kenya; by P. Ganewatte for rural indebtedness in Kagama Kattiyama, a settlement project in Sri Lanka; and by Mead Cain for sale transactions of arable land in Char Gopalpur – a village in Mymensingh District in Bangladesh, and in three Indian villages – Shirapur and Kanzara in Maharashtra and Aurepalle in Andhra Pradesh. Some of the main findings of these studies are summarised in Table 5.1. While taking debts is not as ratchet-like as selling land, the Sri Lanka data are included because of the interest of the comparisons.

The authors would probably be among the first to question a superficial interpretation of the results. In some societies, data on land sales are sensitive, and Polly Hill is sceptical about answers to questions on this subject in Northern Nigeria (1972, pp. 88–9):

> To ask a farmer 'Why did you sell that farm?' is almost as ridiculous as to enquire 'Why are you poor?'. The matter of farm-selling is usually a very painful one to all but the brashest sellers – reasonably enough in a society where sellers are apt to be dubbed 'failures' by others. Although it is often a conjunction of unfavourable circumstances which causes selling, so that a man could fairly retort that he sold a farm because he was so poor, embarrassed informants commonly mention the first contributory cause which enters their heads (providing it is not too painful), or hastily provide the foolish questioner with the kind of simple answer they expect him to be expecting.
>
> (1972, p. 88)

She concludes that in this situation it is impossible to indicate the relative importance of different causes, though she does list the different contributory causes given (*ibid.*, pp. 89–91). Bailey is similarly cautious (1957, p. 62). Such difficulties vary by context. But all the authors carried out careful fieldwork and significance can be attached to the orders of magnitude they identify. Three qualifications must, however, be made.

Table 5.1 Reasons given for sales of land and for outstanding debts

	Bisipara, Orissa, India	Tsakani, Kilifi District, Kenya	Kagama Kattiyama Special Project, Sri Lanka	Char Gopalpur, Mymensingh District, Bangladesh	Three villages (2 in Maharashtra and 1 in Andhra Pradesh) in India
			Percentages		
Year(s) of fieldwork	?1953	1966–7	1971	1976	1980
Number and nature of transactions	57 Sales of land	75 Sales of land and palms	102 Debts[1]	239 Land sale transactions	61 Land sale transactions
Marriage (bridewealth dowry, costs of ceremony)[2]	33	59	21	3	15
Funerals	19	29	20	–	–
Sickness[3]	2	24	7	7	18
Buy food[4]	12	–	10	58	–
Buy plough cattle, bullock	16	–	–	4	–
Other productive investment	–	–	–	8	28
Inconvenience of cultivation, location, or size of plot	4	7	–	1	–

	Bailey (1957, p. 58)	Parkin (1972, p. 59ff.)	Canewatte (1974, p. 13)	Gain (1981, pp. 451-2)
Build or repair house	2	–	–	3
Litigation/compensation/fines	4	2	5	–
Children's education	–	–	–	13
Emigration	2	7	15	–
Other	7	2	40[5]	24
Total	101	130[6]	102	101
Sources	Bailey (1957, p. 58)	Parkin (1972, p. 59ff.)	Canewatte (1974, p. 13)	Gain (1981, pp. 451-2)

Notes:

1 Canewatte's data are for debts outstanding in 1971 and may include debts incurred earlier. (66 farmers who took cultivation loans in 1968/69 and had not repaid earlier. These were, however, scarcely serious loans as only 2 of the original 68 loanees had repaid.) The size of total outstanding debts was (in Rs.): marriage ceremonies – 10 125; purchase of capital goods – 4 392; funeral – 3 380; sickness – 2 600; consumption – 2 550; house repairs – 2 100; pilgrimages – 1 985.

2 In the Orissa village, 15 cases were 'to bring a bride' and 4 'to send a bride'; in Tsakani the cost was bridewealth; in Kagama Kattiyama it was wedding expenses and dowry.

3 The Orissa village had a high incidence of sickness but was isolated and had no close access to allopathic or ayurvedic medicine. Tsakani and Kagama Kattiyama expenditures were mainly or entirely for local (i.e. not allopathic) treatment.

4 In Kagama Kattiyama this was described as 'consumption expenditure'.

5 This includes 15 per cent for pilgrimages and 15 per cent for buying capital goods such as radios.

6 More than one reason could be counted for a sale.

123

First, the reasons for land sales can be expected to vary according to landholding size. This is borne out by the evidence from Gopalpur and the three Indian villages. In the three Indian villages, almost all the sales reported for children's marriage, children's education, and productive investment were made by large owners, constituting 57 per cent of all the land sold, and large owners are recorded as selling over ten times as much land as medium and small owners. The poverty ratchets are to be sought among the sales of small and medium owners, as shown in Table 5.2.

The second qualification is that information about asset disposals is subject to time lags. There are the time lags between fieldwork and the publication of research findings (see pp. 31–2 and 49). Further, Bailey's fieldwork was carried out some thirty years ago, in 1953, Parkin's in 1966–7, and Ganewatte's in 1971. On top of this, the information is retrospective. Cain's data, for example, are for land transactions since inheritance, and the average date of inheritance in Char Gopalpur was 16 years, and in the three Indian villages 23 years, before the respective surveys. Any policy implications that might be drawn must be tempered by recognising these time lags, and the possibility that the proximate causes of impoverishment might have significantly changed in recent years.

The third qualification concerns sequence. Parkin was surprised by the replies he received since funerals and sickness were not cited as frequently as he expected from observations (1972, p. 60). This prompted him to investigate the expenditure during the preceding few years of families which had given 'debt' and 'expedience' as reasons for sales. This revealed that all had had medical and funeral expenses in recent years.

Sequences can be illustrated by two cases. A small and illiterate peasant in Jhagrapur, Bangladesh, is reported to have described thus the process of his impoverishment:

When my father died eleven years ago, I was seven years old. He left my mother and me 11 bighas[9], while there were no other inheritors. My mother could not manage, since there was no-one to do the ploughing and harvesting. So she felt forced to sell part of our land. Moreover, we lost some of our land through extortion.

At present, only 4 bighas are left. Two of these have been mortgaged to two rich peasants. One time we were in great need of food and the other time our hut had burnt down and there was no money to build a new one. . . . If I could pay both of them 70 taka now I would get back the land. But

Table 5.2   Reasons given for sales of land by landholding size

Numbers of land sale transactions

	Char Gopalpur Bangladesh (114 households)				Three Indian Villages (119 households)			
	Land-less	Small	Medium	Large	Land-less	Small	Medium	Large
Children's marriage	–	5	–	1	–	–	–	9
Medical expenses	1	6	2	8	–	–	2	9
Buy food	3	35	44	56				
Buy bullock	–	2	5	3	5	–	4	8
Other productive investments	1	8	3	6				
Litigation expenses	–	3	6		*	*	*	*
Children's education	–	1	–		–	–	1	7
Bribe for employment	–	1	1		*	*	*	*
Pressured to sell by other claimant	–	2	–		*	*	*	*
Other	–	11	5	17	1	4	2	9
Total	**5**	**74**	**66**	**94**	**6**	**4**	**9**	**42**

Source: Cain, 1981, pp. 451–2

Notes

1   In Char Gopalpur, small = 0.01–0.92 acres; medium = 0.93–2.18 acres; large = 2.19+ acres.
In the Indian villages, small = 0.01–3.72 acres; medium = 3.73–13.75 acres; large = 13.76+ acres.

2   * = not a category in the original table.

3   For further information to throw light on the many intriguing questions raised by this table, the reader is referred to Cain's original article.

> where do I get so much money? And moreover, they do not
> like to give it back before the whole period has passed.
>   (Arens and van Beurden 1977, p. 14)

By this account, the physical weakness of the family, extortion,
shortage of food, a hut burning down, the lack of a small sum to
redeem a mortgage, and the assumption that rich peasants would
object to early repayment, have made the peasant poorer and keep
him poor.

Another sequence comes from the life of Hem Nath Brahmin
in Nepal.

> Originating from a hill village in Gulmi district where he
> owned a small parcel of land, he was obliged in 1971 to
> emigrate to maintain his wife and three children. He sold
> the land and came to the terai (lowland) to buy land there. He
> was unable to find an adequate plot with the small sum at his
> disposal and set up a tea shop in the meanwhile to keep the
> family alive. The tiny business survived for around three
> years, but he was obliged to extend credit to maintain a
> clientele and when he fell sick with cholera four years ago the
> accumulated debts of between Rs. 700/- and 800/- could not
> be recouped and the hospital charges led to his financial
> collapse. His wife returned to the hills with their youngest
> son of five to live with his wife's brother, and his other son of
> thirteen had to be sent to stay with his mother's brother in
> India. The only member of his family remaining with him is
> his ten year old daughter. 'I have no land and no place in my
> village now; I am sick and can barely work. I may not see my
> family again.'
>   (Blaikie, Cameron and Seddon 1979, p. 47)

Here the sequence appears to have been pressure of population on
land forcing emigration, too little money to buy land, the
prevailing poverty of others which forced Hem Nath Brahmin to
give credit from his tea shop, and then sickness and hospital care
leading to the ruin of his business, and the dispersal in
desperation of his family, split into three for survival.

The six cases, five area studies, and now these two sequences
together with other evidence, provide the basis for some
reflections.

There is a danger of over-generalising. In the first draft of this
chapter, before I was aware of Cain's work, I wrote 'Social
expenditures — dowry, brideprice, weddings, and funerals — are
strikingly dominant as reasons given for selling land or taking

debts . . .'. However, in the three Indian villages not one of the small landholders reported selling land for a marriage. The contrast between Char Gopalpur, where so many land transactions were reported to buy food, and the Indian villages where there were so few, is also revealing, and can be attributed substantially to the relief programmes mounted in the Indian villages when there was serious drought.

Ratchets are not all in the direction of deeper poverty. In Char Gopalpur, it is true, the proportion of landless increased from 20 per cent at inheritance to 29 per cent in late 1976. But in contrast, the movement was in the other direction in the three Indian villages: the decline in the proportion of landless was from 32 to 12 per cent in Aurepalle and from 41 to 18 per cent in Kanzara. Among other factors this is attributed by Cain to effective public works relief in bad years, to land reform legislation which discouraged large owners from accumulating more land and which induced them to sell to tenants, and to opportunities for larger owners to invest in more intensive cultivation through irrigation.

Sickness emerges as a common cause and form of poverty ratchet. The very low incidence in Bailey's study of Orissa is misleading. The climate was considered unhealthy and there was always sickness in the village; but the local doctor (*baidyo*), the Brahmin, and the diviners were called in when there was sickness and did what they could with traditional measures. The doctor expected a fee of a glass or two of rice only. The Brahmins and the diviners would haggle and hold out for more, but it seems that the total cost of treatment was very low, that a certain fatalism prevailed about sickness and that visits to the ayurvedic or allopathic doctor were rare (Bailey, 1957, pp. 18–19, 113–14). The village appears to have been at a stage before the higher costs of ayurvedic and allopathic medicine had begun to bite. In Tsakani in Kilifi District, sickness was listed as a factor in land and palm sales in 24 per cent of the cases reported. In Kagama Kattiyawa in Sri Lanka, the lower figure of 7 per cent of debts incurred because of sickness may be a consequence of the virtual absence of malaria from Sri Lanka at the time of the fieldwork, the free rice ration, and the extensive and cheap health service.

Sickness occurs often in the six case studies and the two sequences. In three of the six case studies, it was the cost of medical treatment (for Don Pedro's surgery in hospital, for Tiyo Oyo's hospital treatment and dextrose, for setting the broken leg of the Bangladeshi boy) rather than the sickness or accident itself which led to the main disaster (selling the mule, selling the carabao, the boy being lame and dependent for life because the money could not be raised). The plight of Hem Nath Brahmin in Nepal illustrates the multiple effects of sickness on the vulnerable. His cholera dealt him a treble blow: there was the sickness itself which put him in hospital, weakened him, and stopped him earning; then the costs of treatment; and finally his inability, sick as he was, to recoup debts owed to him – all combining to ruin his tiny tea shop business and scatter and virtually destroy his family.

In all these instances, those whose treatment was expensive were males. It is a question for research to what extent there are differential costs and types of treatment for men and women in different social conditions. In a village in Karnataka in South India, Sudha Rao reports that almost all those receiving treatment for TB were men. Women may often simply have to bear their sickness, or may receive lower cost treatment; and may thus less frequently precipitate poverty ratchets for the cost of treatment.

Many poverty ratchets are made worse by urgency. Contingencies cry out for action. Assets have to be disposed of in a buyer's market. The seller, or the pleader for a service or loan, is known to be desperate and up against time. Raising resources for bridewealth or dowry can sometimes be spun out. But funerals cannot be delayed; sick people must be treated; a broken leg must (one would suppose) be set; a new hut must be built when the old one has burnt down; food must be obtained to relieve hunger so that work can be done and for very survival.

> If an ox drops dead in the middle of the ploughing season, when no-one wants to lend or hire their cattle, and if the peasant must sell land to buy a new ox, he is in no position to drive a hard bargain. In general the market of buyers is restricted to his own village, where everyone knows his predicament. The same can be said of the cost of mortuary rites which must be concluded within at the most twelve days. There is not the same immediacy about sales to provide for a marriage or to build a house, but the price is still kept low because the prospective buyers know that the time which the seller can spend bargaining is limited.
>
> (Bailey, 1957, p. 59)

It is not surprising then to find distress sales and distress prices. The Bangladesh mat-weaver appears to be an exception: he sold his 0.19 acres to his brother for 2 000 taka, a good price presumably given because of kinship. But Don Pedro got only about two-thirds of what he felt his mule was worth and Tiya

Teria less than half what was considered the value of the Sumagaysay's buffalo; and the Nigerian land, as is usual in famines, was sold at very low prices. Poor people cannot wait. Because they cannot wait, they get low prices. Ergo, they get low prices because they are poor.

Vulnerability to poverty ratchets is heightened when the assets to be disposed of are big, indivisible, and productive. This applies to both land and animals. Bailey (1957, p. 60) gives an example of indivisibility of land. A field may be the smallest unit which can be sold, but a man's smallest field may be worth more than the money he needs. If the land is his only capital resource, the buyer is in a strong position and the seller is likely to get no more than he needs even though the field is worth more. Similarly, large livestock present problems. The Sumagaysays sold their carabao for P330, but the costs of hospital treatment and medicine were only P250, not including food. It seems probable that even with the low price they got for the animal, they raised more money than they needed simply because it was 'lumpy' and could not be divided. It is also possible that the small farmer who bought the carabao had some sense of how much money the Sumagaysays needed, and kept the price down to near that. It is here that small stock are better for the very poor. They can be realised in smaller lumps. Sales of sheep and goats and the purchase of grain with the proceeds are a widespread defence against the food shortages of the hungry season (O'Leary, 1980), even though they may fetch far lower prices at such times (Hill, 1972, p. 164).

But whether the sale is of land or livestock, and whether these are large or small, a productive asset is being disposed of, and, with the exception of old livestock past breeding, this reduces the subsequent food supply and income of the family. Without their mule, the Martinez family lost firewood they could sell and earnings hauling crates of plums, besides having to put in more work to bring home their own harvest. A day's labour by Tiyo Oyo had been worth P10 with the carabao; without it he could earn only P6. The same applies to disposals of land, when small farmers become landless and have only their labour to sell. The Bangladeshi mat weaver who previously grew his own cane and reeds had, after losing his land, to buy them. The loss is also greater when the asset is appreciating. The Italian peasant's hog was no doubt being fattened and increasing in value. Small stock breed fast and provide a potentially rapid means of gaining. But the dice are, as always, loaded against the poor. In a village in the Dominican Republic:

Most day laborers have neither the land to support cows or pigs nor the capital to buy them. More seriously, day laborers rarely can hold animals long enough to fatten them: with their continual money shortage, serious illnesses are always forcing quick sales for the cash needed to pay doctors and buy medicine.

> (Sharpe, 1977, p. 46)

The less people have, the more vulnerable they are, and the harder it is for them to rise. A study of peasant farmers in Northern Nigeria over four years found that those with both livestock and land retained virtually the same acreage, but those with no livestock ended with 12 per cent less land (Simmons, 1981). Sales of livestock were, it seems, a buffer against sales of land. N. S. Jodha, comparing several studies of drought years in India and their sequels, found that the decline in assets was greater and the recovery was generally slower in the case of small farmers compared with large (1978, p. A41).

There is, too, the pervasive trap of indebtedness and high interest rates. While these may not be mentioned as contingencies forcing poverty ratchets, they are often a major factor building up intolerable pressures to mortgage an asset or to surrender an asset that has been mortgaged. The cruellest cut is where the less educated and the poorer people pay higher rates than others who are better off. Michael Howes (1980) found a not very numerate peasant in Thailand charged 120 per cent interest on a 2 months loan (an annual rate of 720 per cent). Margaret Haswell found in Genieri village in the Gambia that during the 1973–4 agricultural season, short-term loans repayable at harvest bore interest rates ranging from 49 per cent of 157 per cent for the eight months of the loan; and that the lowest rates applied to those who had one or two head of cattle which could be sold in repayment if necessary, and the highest rates to those with no assets in livestock (1975, p. 186).

With similar perversity, rates of interest rise in drought years (Jodha, 1978, p. A46). High interest rates, or bad credit-worthiness which prevents loans, make it even more likely that the poor will be forced to dispose of whatever assets they have. There are indeed what Jodha calls 'asset depletion-replenishment cycles' (1978, p. A38). But those best able to replenish are those least depleted. In the words of the Bible, 'For unto every one that hath shall be given, and he shall have abundance: but from him that hath not shall be taken away even that which he hath' (St Matthew XXV, verse 29). Depletion heightens vulnerability. The depleted household has fewer buffers against contingencies. One ratchet effect leads to another. For many of the poor, depletion makes them permanently poorer, and permanently vulnerable to becoming yet poorer still.

Powerlessness

There is a peculiar obviousness, almost a tautology, about the links between powerlessness and poverty. What is most important is clear and well known, and yet so discomforting for the powerful that we continue somehow to overlook it and talk about other things.

So let us start with the obvious. At the local level, those who are powerful are often described as an elite. They are at the opposite pole to the poor along each of the dimensions of deprivation: they are relatively well-off in assets and income; they are physically strong (healthier, with larger families, larger bodies); they are secure – able to weather disasters and to obtain medical treatment when sick; they are spatially, socially and politically at the local centre of things – well-informed, able to educate their children, and able to draw on government resources and the machinery of the State. Their power derives from these factors, from their solidarity as a class, and from the powerlessness of the poor. In an earlier age, such rural elites were often seen as benevolent; today they are more often regarded as exploitative. And this shift of view within limits reflects a change in reality.

Exploitation of the powerless poor by the local elite takes many forms, but three clusters stand out: nets, robbery, and bargaining and its absence.

i) *Nets*

Local elites stand as nets between the poorer people and the outside world, in the sense that they catch and trap resources and benefits. Most government, parastatal and private sector programmes and campaigns are either designed intentionally for the elites, or so designed and implemented that they are likely to be intercepted by them. It is a notorious commonplace how, almost everywhere in the third world, credit and marketing cooperatives have been dominated by the larger farmers who have used them for their own benefit, at the cost of smaller producers; how agricultural extension staff are locked in with the more 'progressive' farmers, and with men rather than women; how tube-wells, tractors, irrigation water, subsidised fertilisers, and cheap credit are obtained by the larger farmers; and how the benefits of the green revolution have been unequally distributed.

The eloquent case study The Net (BRAC, 1980) (see pp. 69–70) lays bare the power structure in ten villages in Bangladesh. This may be an extreme case. All the evidence I have seen suggests that Bangladesh presents particularly bad conditions of rural exploitation, and the Bangladesh Rural Advancement Committee may have chosen an area where power was much abused, even by Bangladesh standards: the villages were near the border where there had been refugee movements, where poor and despised Adivasis (tribal people) lived, where the Army was present, and where there was a frontier of communal forest resources inviting expropriation by the powerful. Even so, similar conditions are found in other places and other countries and the findings of the study present an agenda of questions to be asked elsewhere. The BRAC study found, for example, that food relief was intercepted by the powerful, and over one period only 24 per cent of the food sanctioned for food-for-work had been distributed, the remainder having 'disappeared' (BRAC, 1980, p. 78). Large sums of money accrued to a small group of lower and middle level politicians and officials and local wealthy men through appropriation of relief food and subsidised food and its resale. The main effect of food programmes destined for the poorest was to enrich those who were less poor.

It is not just that the powerful intercept; it is also that the price demanded for passing on benefits can exclude many of the poor, quite apart from diminishing the value of the benefits. One of Jansen's informants said he tried to get loans from banks at the thana headquarters, but he failed because the bank personnel demanded too much in bribes (1978, p. 49). Epstein was told by Malla, a member of a Scheduled Caste in South India, that he was aware of the favourable loan arrangements for members of the Scheduled Castes, but that to qualify for one he would need the signature of an official who he knew would not sign without a bribe of Rs. 150. 'Malla went on jokingly: "If I had Rs. 150 ready to give in bribes I would not need a loan at all!" ' (Epstein, 1973, p. 161)

Nor does this exclusion apply only to loans. We have already seen (p. 120) treatment for a broken leg withheld because the sum illegally demanded could not be raised. To make things even worse, the rich and powerful may receive free the services for which the poorer have to pay.

ii) Robbery

The elite is also well placed to use deception, blackmail and violence to rob the poor. Police, government officials and the larger landowners and traders have common interests and understandings. Illiterate people are induced to sign documents they do not understand or which are falsely described to them; ignorant, misled, and fearful, they unwittingly renounce their rights in land, accept debts without knowing the terms of repayment, mortgage possessions without being clear about how they can retrieve them, and accept terms for loans which are grossly extortionate. They then lack recourse to justice – since they do not know the law, cannot afford legal help, fear to offend the patrons on whom they depend, and would anyway be bringing their case before members of the very elite against which the appeal would lie. If they complain or resist, they can be brought to heel by a visit from the police, a threat of prosecution or arrest, the calling in of a debt, a refusal of employment, or violence – the burning of a hut, a roughing up by thugs, or worse. Open violence against the poor may be the last sanction not often needed, but it can represent the visible tip of an iceberg of very widespread subtle and not so subtle intimidation and the fear it creates. Violence against vulnerable groups in rural India is recorded almost daily in the Indian press, to the credit of its courageous journalists; and much more would be reported in other countries were their press as free as it is in India.

Some of the worst situations for the poor are where military, police and a local elite together exploit small farmers and the landless, especially when the latter are politically weak refugees (Chambers, 1979). In South Kivu in the mid-1970s the Zaïrean Army was arbitrarily seizing fish and small stock of Barundi refugees, and forcing young men to carry for them in the mountains. Some of the refugees never returned. For Bangladesh, *The Net* analyses a similar situation concerning a weak tribal people, intermittently refugees, who were being persecuted and expropriated by an alliance of elite, army and police. Let it speak for itself with one incident, by no means the worst:

> Not only are powerful people able to defy the law without suffering any consequences, because of their good links with the Police and BDR (Bangladesh Rifles) they are able to use their power to make personal profits. For example in 1978 Rafiq Daktar was passing the house of a poor Koch [Adivasi or Tribal] when he was attacked by a buffalo. Rafiq Daktar ran off and accused the Koch of owning a dangerous mad buffalo,

telling him to sell it immediately or he would inform the Police. When the man refused he went to the Police Station and returned with two Policemen, who threatened him with arrest. The man became afraid and agreed to sell the buffalo to Rafiq Daktar for Tk. 1 200/- (the market price was Tk. 2 500/- to Tk. 3 000/-). After buying the buffalo Rafiq Daktar slaughtered it and sold the meat to other people giving a share to the Police and BDR. He never paid the price, however, and when it is demanded he replies that most of the meat was taken by the BDR and Police, who did not pay, so why should he pay? The local people have demanded an account from him, but he has not given it.

(BRAC, 1980, p. 98)

In cases like this, the slightest presumption, justified or not, of legal or moral error on the part of the poor can be turned against them for profit to the rich; and appeals for justice can simply be shrugged off.

iii) Bargaining and its absence

Unequal power is reflected too in bargaining and its absence. This applies to prices paid in distress sales, as we have seen; to low payments when assets are pawned or mortgaged, even though the assets are likely to be lost through non-repayment; to reluctance of those who have mortgaged assets to try to redeem them for fear of prejudicing future loans, and for fear that the creditor would anyway refuse; and most common of all, to the prices paid for the labour of the landless.

In many places, the landless face the harsh arithmetic of supply and demand. In the words of a Filipino (Ledesma 1977, p. 27) 'As a landless worker, it is solely your body that earns a living'. Where the bodies exceed the work available, the price paid for them goes down, or work and payment are shared, as practised in parts of South India. Declining real wages are far from universal, but are widely reported for parts of Asia (e.g. ILO, 1977). Employers of casual agricultural labour, moreover, switch from payments in kind to payments in cash and back again, adopting whichever makes labour cheaper. But what is even more significant, and more general, are the differentials within populations of rural labourers. What happens to those groups who are most disadvantaged is easily lost in statistical averages. But those who are weaker socially, physically and politically are least able to bargain and get paid least.

The clearest group discrimination is against women. Almost everywhere, the earning power of rural women is less than that of rural men. Sometimes men are paid more than women for the same work. More commonly, tasks are segregated by gender. Leela Gulati records that in brickmaking in Kerala, women are confined to the strenuous work of carting bricks and never earn more than Rs. 5 a day, whereas men can expect to make Rs. 10 or more (1981, pp. 40–44). Again, in the coir industry, women husk-beaters make less than Rs. 4 per day whereas men who transport husks make between Rs. 12 and Rs. 16 (1981, p. 148). The weak bargaining position of women has several dimensions – low social status, the fact of male physical domination, and the strong maternal drive for work. Women with children to feed and no food cannot bargain, except desperately for some bare minimum. And they are least likely to negotiate anyway at the bad times of the year, especially during the rains when food is short before harvest. Thus, again in Kerala, Joan Mencher records interviewing groups of women working in the fields in the pouring rain who said that they had no idea what they would be paid for their day's work, but that they had no choice but to work. As one woman put it:

> This Maharaja called us. What can we do? Can we sit at home and listen to the cries of hunger from our children? For one or two nights we can bear it, but then we will come to whoever offers us some work. This is a bad time and the children suffer so much. Even our own hunger is not easy to bear.
> (1980, p. 1799)

Those who are disabled, physically weak, or destitute are also unlikely to bargain. One of Jansen's informants in Bangladesh, suffering from chronic asthma, said the pay he received was determined by the landowner. 'When asked if he ever asked for higher wages, he replied that he is afraid to ask for more pay – besides, he said, what landowner would seriously listen to his request' (Jansen, 1978, p. 28). For those who are destitute, seeking work is making a distress sale of the only asset that remains to them, their labour. Their powerlessness to bargain is manifest in their clothes, bodies, and demeanour. A Bangladeshi village woman whose grown daughter was sick had to get work each day for them to survive: a day's earnings was three meals – one only each day for the daughter, and two for the mother to give her strength to work.

Willie Henderson has described (1980, pp. 231–2) two unmarried sisters in a village in Botswana who were in a similar plight. On one occasion they and their eight children were sitting in the yard crying for hunger. They earned money collecting buckets of water. The round trip was nearly two miles. The normal price for a bucket was five cents or ten cents; but to earn ten cents, *they* had to fetch six buckets. They had complained to the headman but he had done nothing. '...the fact that these women were known to be destitute or virtually destitute meant that other families tended to exploit them.' People in such straits have nothing with which to bargain; they can only plead and beg for work and food and accept whatever they are given.

Powerlessness in labour relations is often acute for migrant labourers. If village labourers become desperate and migrate, this may be taken as an excuse by their village patrons to withdraw from their obligations to provide support, while those who employ migrant labourers accept no responsibility. Health then is critical. Sick labourers get neither work nor help. Jan Breman has described the annual migration of poor people (Dublas) from a south Gujarat village in India to get seasonal work in brickyards:

> The continual alternation of work situations... means complete subjection to a labor system in which the employer, owing to the temporary nature of the engagements, rejects any obligation to provide for the most elementary needs of his employees. This sometimes leads to their being treated inhumanely. During my stay (in the village) for instance, a grandfather, a father, and two daughters returned to the village in mid-season. They suffered from typhoid fever and had been sent away from the brickyard so as not to infect the other laborers. Gravely ill and penniless, they had found their way back, travelling surreptitiously by train and getting a lift in an oxcart. Once they were in Chikhligam, not one of the farmers for whom members of this household sometimes worked gave them any aid. I found these people in their hut, uncared for and without food, lying on jute bags. Two of them died of the fever the next day.
> (1979, p. 252)

Examples such as these present only one part of the whole. There is another side: reciprocal relations between the powerful and the weak, a sense of obligation within a community to employ landless labourers (Hayami, 1978, p. 29), and sharing and mutual help between kin and between the poorer people who are often so generous themselves with the little they have. However much it may be brushed aside by those who see, or wish to see, only the selfish exploitation of power, there is also generosity, altruism and an ethic of sharing. The great religions enjoin charity; Christ said 'The last shall be first'. A desperate household, the wife sick, the husband unemployed,

> was surviving on a tiny amount of savings and on a loan of food stuffs worth about Rs. 40/- from a shopkeeper who had employed him [the husband] previously. No interest was to be paid on this loan, nor was there a repayment date – 'it is just help, and without such help we would not be able to go on'.
> (Blaikie, Cameron and Seddon, 1978, p. 48)

Nor are elites, classes or bureaucracies monolithic; nor are they all parasitic and corrupt.

But evidence of the directions and magnitudes of change is depressing. In some earlier societies, and some which still survive, the sharing and reciprocity of what is, ironically, termed *primitive* exchange (Sahlins, 1974, p. 210ff.) reduced inequalities and hardships by redistributing gains, assets, and food within the community, and inhibiting economic differentiation. However, the traditional responsibilities of the rich, or less poor, for those who are poorer, have been fading: the storage of food by the wealthy to be provided to all in bad times has withered in Nigeria as in West Bengal; communal rights to land, water, grazing, forests and fishing have been appropriated by the stronger families and households, or by powerful outsiders, and have become private and exclusive; households once differentiated more in terms of their labour power and their ability to use their labour to make communal resources productive, are increasingly differentiated by wealth, education, and command over external material resources and political power.

Sometimes, in these processes, all are better off, though some much more than others; quite often many find themselves worse off. Traditional supports and coping mechanisms are weakened or removed, and new ones cannot be found or improvised to take their place. Despite variations, these are widespread trends. Economic growth which gives also takes away. Earlier observers of rural life were right in their time, in the prominence they gave to reciprocity and sharing: they were more common. And observers today are also right in giving more prominence to the acquisitiveness of the smaller family: it is already more significant than many yet realise.

Of course, this is not the whole story. There are many places where the introduction of cash crops, irrigation, input packages, and infrastructure have generated prosperity, more employment, higher real wages, and a better livelihood for all except the most indigent and unfortunate. The directions of change vary – by country, by region, by social group, and by gender. If these directions have been positive for many, there remain hundreds of millions for whom they are negative. It is these people who present the awfulness, the tragedy, and the challenge. For them, poverty, physical weakness, isolation, vulnerability and power-lessness become more and more tightly integrated. They are more and more trapped in deprivation. Stopping the downward slide, let alone reversing it, becomes for them more and more difficult. There is the imagery of games: the dice loaded against the poor: snakes and ladders where there are more snakes than ladders at the bottom of the board, keeping the poor poor and making them poorer, and more ladders than snakes near the top, keeping the rich rich and helping them to get richer. Or there is the trap closed on the rural poor, with the less poor – rural, urban and rich country alike – sitting on the lid. The problem is to reverse the trend – to load the dice differently, to make more ladders near the bottom, to shift the privileged so that the lid can be lifted. It is this challenge that the rest of this book will try to address.

Notes

1   It is not only the non-rural outsider, of course, who may despise what poor people say. The better-off rural people may do so even more. In a meeting with Somali cattle owners once, an old man who seemed to me to be speaking a lot of sense was openly laughed at by others present. When I later enquired why, I was told it was because he had hardly any cattle.

2   Which words I, like so many others, was taught to sing at an early age. It is a shame that this verse disfigures what is otherwise a simple hymn of wonder.

3   In descending order the percentages of poverty cynicism in the whole population surveyed were:

United Kingdom	27
Germany	17
Belgium } France }	13
Denmark } Netherlands }	11
Italy	9
Ireland	8
Luxembourg	7

(CEC, 1977, p. 88)

Compared with others, the poverty cynic group 'is older, less well-educated and not so well off'. However members are not unhappy with the life they lead. They tend to put themselves fairly high up on the rich/poor scale and to the right of the political spectrum (*ibid.*, p. 80).

4 For micro-detail in trying to make progress with this subject I have found the following works especially useful: Arens and van Beurden, 1977; BRAC, 1979, 1980; Bailey, 1957; Blaikie, Cameron, and Seddon, 1979; Breman, 1979; Dolci, 1966; Epstein, 1973; Gulati, 1981; Haswell, 1975; Henderson, 1980; Hill, 1972 and 1977; Jansen, 1978; Ledesma, 1977; Lewis, 1959; Moore and Wickremesinghe, 1980; O'Leary, 1980; Parkin, 1972; and Srinivas, 1976. I have not tried to penetrate the mode of production debate. For advice on that subject, without which I might never have managed to write this book, I am grateful to John Harriss.

5 For a revealing analysis of India village survey data which identifies two polar types of villages – those which are remote, with less irrigation, and more equal distribution of assets, and those which are more accessible, with more irrigation, and greater inequalities of wealth, see Dasgupta, 1975.

6 The more correct but less well known term, according to Douglas Thornton, is pawl.

7 For the death of a goat, see Gulati, 1981, pp. 56–7.

8 But see Bailey 1967, Parkin 1972, and Cain 1981, and pp. 122–3 and 125 below.

9 In Bangladesh, a bigha is about 1/5th of an acre. This is a regional figure; in Bihar the bigha is bigger.

CHAPTER SIX

Seeing what to do

The philosophers have *interpreted* the world in various ways; the point however is to *change* it.
Karl Marx, 1845 (his emphases)

You see things, and say why? But I dream things that never were, and I say why not?
Robert S. McNamara, quoting George Bernard Shaw, *Back to Methuselah*

In trying to see what to do, outsiders' unavoidable paternalism can be offset in two ways: first, by starting with the priorities and strategies of the rural poor themselves, noting that though some are 'foxes' with a varied repertoire of petty activities and others 'hedgehogs' locked into one relationship, all share the aim of a secure and decent livelihood; and second, by concentrating on what outsiders and the rural poor agree in saying no to. Rural development can be redefined to include enabling poor rural women and men to demand and control more of the benefits of development. In practice, outsiders tend to think of interventions through plans, projects and programmes implemented by government field organisations, but this neglects two powerful mechanisms: prices; and the law. Analysis to see what to do can usefully include assessing costs and choices, examining causes and constraints, and finding and making opportunities, especially ways for poorer people to gain control over assets at the critical time when they increase in value. The brick wall of political will is best tackled through practical political economy which takes account of who will gain and who will lose. Some feasible actions can be found where the poor gain and the powerful also gain or do not become worse off. Elites prefer actions which tackle physical weakness directly, but poverty and powerlessness are more basic. To tackle them directly and effectively often requires enabling the poor to mobilise and organise, to demand, gain and maintain control over assets and income.

Whose priorities?

In trying to see what to do, non-rural outsiders are trapped by
core-periphery perception and thinking. Looking outwards and
downwards towards the remote and powerless, their vision is
blurred. They see most clearly what is close by; they see action
starting from where they are. The very words reflect the problem:
'remote' means remote from urban and administrative centres,
from where most of the outsiders are; and 'what to do' implies
initiatives taken by them in the centres of power. However much
the rhetoric changes to 'participation', 'participatory research',
'community involvement' and the like, at the end of the day there
is still an outsider seeking to change things. Marxist, socialist,
capitalist, Muslim, Christian, Hindu, Buddhist, humanist, male,
female, young, old, national, foreigner, black, brown, white –*who*
the outsider is may change but the relation is the same. A stronger
person wants to change things for a person who is weaker.

From this paternal trap there is no complete escape. A
decision not to act is itself an action. A person who withdraws or
who abstains from intervening, is by that withdrawal or
abstention still intervening by default. The weaker person is
affected by what does not happen but which might have
happened. There is, however, a partial remedy. Respect for the
poor and what they want offsets paternalism. The reversal this
implies is that outsiders should start not with their own priorities
but with those of the poor, although however much self-insight
they have, outsiders will still project their own values and
priorities. In what follows, I too am trapped, an outsider asking
what poor people want. All one can hope is that the effort of trying
to find out, of asking again and again and doubting the outcomes,
will check some of the worse effects of core-periphery
paternalism, and that the more the priorities of the poor are
known, the easier it will be to see what it is best to do.

Priorities and strategies of the poor

For those who are neither rural nor poor to know the priorities of
those who are both is not as easy as it sounds. The rural poor are
dispersed, isolated, uncommunicative, rarely asked their views,
frequently masked by others, selectively perceived, deferential.
The silent cannot be heard. Direct approaches distort impressions: replies in interviews notoriously mislead, especially when
respondents believe that their replies may bring benefits. An
indirect approach may help, drawing on social science research,
especially case studies of social anthropologists and social

workers, and agricultural economists' understanding of the behaviour of poor farmers. On the basis of such evidence something can be said about what poor people want, inferring their priorities from what they do as much as, or even more than, from what they say.

Differences must be noted. Environments and conditions vary – from those where land is scarce (many in Asia), to those dominated by large landholdings (many in South America), and those with adequate land but low labour productivity (more common in Africa).[1] Nor are poor rural people a uniform mass; nor are their strategies all the same. Even in the same locality, there can be a big contrast between the strategies of those with some land and those who are landless.

All, however, tend to share the characteristics described in the last chapter, being, besides poor, also physically weak, isolated, vulnerable, and powerless. Survival is a constant preoccupation, and at its most basic survival means food and not being sick or injured. Food is a continuous priority, with the well-known consequence that subsistence farmers and other poor people are averse to risks and cautious about new practices which might jeopardise their familiar sources of food or make them poorer. Health is a priority whenever sickness or injury occur, but is often harder to control than the food supply. The struggle is both a daily one for basic necessities, and also a longer-term one to gain control of assets and build up buffers against contingencies. Many follow the strategy of the large family – as insurance against children dying, to provide for parents in old age, and because children help the household early in life (caring for siblings, looking after small stock or calves, running errands, collecting firewood, water, fodder and dung, scaring birds, sowing seeds, removing stones from fields and ticks from animals, and the like).

Beyond these common characteristics, two dominant strategies stand out. A proverb of the Greek, Archilochus, says 'The fox knows many things, but the hedgehog knows one big thing' (Berlin, 1953). Poor rural people separate into foxes and hedgehogs.

The foxes are those who contrive a living from a repertoire of different petty enterprises and activities, which may include small-scale farming. For many, seasonal migration to fill agricultural slack periods is a regular, if often desperate, measure taken by some or all of the household. Wherever they are their enterprises and activities have low productivity and bring low returns: growing subsistence crops and vegetables on whatever land can be found; casual labour during the agricultural season; domestic work and odd jobs; temporary employment on public works; fishing, hunting, gathering, gleaning and scavenging; keeping smallstock either for others, or on their own account – goats, sheep, pigs, hens, ducks, pigeons, turkeys, rabbits; bee-keeping; using their own or a borrowed or hired mule, donkey, camel, bullock and cart, bicycle rickshaw, bicycle, or handcart to transport people, crops, vegetables, fish, commercial goods, diesel, water; cutting, collecting and selling fodder, wild insects, fruits and plants, medicinal herbs and roots, grass or reeds for thatching, wood for fuel or building; making and selling charcoal, twine, leather goods, nets, mats, bricks, pots, or tiles; petty hawking; craft work; blacksmithing, carpentry, building, thatching; begging; and theft. For part of Northern Nigeria, Polly Hill writes of most of the poorest men who own little or no land that they devote much time to:

> ... looking for local work and contacting people; they are never fully employed for any length of time; and they commonly eke out a living by following a number of different occupations such as farm or general labouring, the collection and transportation of produce and manure, and the making of cornstalk beds – unlike richer men, some of whom are part-timers traders in the city, their work is mainly locally-based. ... The multiplicity of ill-paid, part-time occupations (most of them of a petty nature) which are open to such men, ensures (I think) that few of them become a burden on the community as a whole, unless they are ill or old and lack close relatives who help them. ...
> (1977, p. 103)

In most rural societies, poor women as well as poor men turn their hands to many things: they have to, to survive.

In contrast, the hedgehogs are those who have only one enterprise or activity. They include some subsistence farmers, and some single-species pastoralists; outworkers for a single urban-based business, like full-time weavers in rural India; and most clearly, those tied by obligations to working for one person or family, like those of the San in Botswana who have become dependent, or labourers in South Asia who are bonded through debt to work for one master. The master may provide the basic security for a survival strategy, but little more. At the one master's beck and call, bonded labourers cannot diversify sources of food and income. They achieve a certain low-level security, but have no escape. They are locked to a single support.

The strategies of the poor can also be seen in terms of the three options of Albert Hirschman's *Exit, Voice, and Loyalty*

(1970). Exit is a strategy for some, either through migration or through educating children in the hope that they will get well-paid employment elsewhere. Voice, in the form of organisation, protest and collective negotiation or force, is not common, but becoming less rare at least in South Asia. Loyalty, in the sense of accommodation to and working within the local society, is by far the most common strategy. Exit and voice are risky, and poor people abhor risks. They prefer familiar subsistence and security. When the West Bengal Government first tried to register sharecroppers so that they would hand over only the legal maximum of one quarter of the crop to their landlords instead of the customary one half, some sharecroppers pleaded 'with tears in their eyes' not to be registered. They feared the insecurity which would follow if they alienated their landlords – the loss of loans to meet contingencies such as when food ran out, a child was sick, a dowry had to be found, or a ceremony had to be performed. Could the government, whatever promises were made, be relied on for this help? What government, anyway, would give credit for dowries and ceremonies?

Such questions pose ethical dilemmas for outsiders. The best approach in each situation may be an unconstrained dialogue with the poor, and an effort to learn from them what their priorities are. What they say will vary – between individuals, between men and women, and between households, occupations, communities, societies, ecological zones and countries. Their priorities may be land, animals, irrigation, tools, seeds, markets, good prices, water for drinking and washing, roads, a school, a temple or shrine, a cinema, roads, defence against raids or vermin, firewood, veterinary services, basic goods at fair prices, employment, minimum wages, credit, protection against landlords and against extortionate interest rates, or many other things. But this is an outsider's list. The priorities of the poor will often surprise outsiders, and those of the poorer will often differ from those of the less poor.

For survival, food and health come first. Or taking the five dimensions of deprivation, poor people may put first reducing poverty, or physical weakness, or isolation, or vulnerability, or powerlessness. But these are again an outsider's categories. When landless people in Bangladesh were asked their priorities, the answers were clear. Their proposals 'were all inclined towards activities which would generate incomes. They talked often of aid coming in the form of inputs for productive and trading activities rather than as ready consumables' (BRAC, 1979, p. 26). It is no coincidence that poverty is the most commonly mentioned dimension of deprivation, and used as a proxy for the others. In

Bangladesh as elsewhere, it seems that most poor rural people seek first an adequate independent source of food and income under their control; or in other words, a secure and decent livelihood.

Objectives for outsiders

Objectives for outsiders can, then, be expressed as a reversal, putting first the wishes of the poor themselves. But this cannot be all. Dilemmas remain: from conflicting values and objectives; from times when outsiders' knowledge is believed to be more valid than rural people's knowledge for achieving what poor people want; from trade-offs between short and long-term costs and benefits; and from outsiders' need to be true to themselves. The question whether to give medical treatment against a patient's wishes, but in order to save her or his life, is an example of the more general problem of power and paternalism. I see no universal solution to this. But for practical purposes in rural development, a partial answer is to concentrate on those aspects of life where outsiders and the rural poor agree. Peter Berger ends his book *Pyramids of Sacrifice* with an appeal for people of different ideologies to find common ground by looking at specific situations to which there will be a common *no*. Outsiders and the rural poor may agree in saying *no* to children dying, to preventable disease, to famine, to the poor becoming poorer, to exploitation of the poor by the rich. Agreement on points such as these can provide a moral foundation for the next steps, to see what outsiders should do.

But outsiders think they know best. Some will say that the rural poor do not know what is in their interests; or that with greater awareness (which is liable to mean by agreeing with the outsider) they would have other priorities; or that they should confront their powerlessness by organising against their rich exploiters; or that they should be encouraged to have longer time horizons; or that they must be enabled to see what they would want if they knew what they really wanted. But if vulnerable people have short time horizons who is justified in imposing long ones on them? If they have low risk strategies, who is justified in thrusting upon them strategies with high risks? It is safer and more humane to proceed by short steps into what can be foreseen than by long leaps into the unknown, in the meantime gaining experience on the way. Changing power relations and the distribution of wealth may often be a necessary condition for major improvement. But any strategy towards that goal which in

145

the short-term makes life appear worse for those who are poorer needs very, very careful scrutiny.

There is no complete escape from the way outsiders project their ideologies and values into analysis and prescription, but at least we have identified two antidotes: first, repeatedly to enquire and reflect upon what poor people themselves want; and second, to return again and again to examples of the unacceptable, and to analyse these rather than theoretical abstractions. A continuous enterprise of seeking to learn from the rural poor and of exercising imagination in seeing what to do is one way of setting directions and correcting course. Without this, outsiders' interventions are all too easily propelled by paternalism in directions which leave people worse off in their own eyes than they were before.

To stress livelihood can, then, only be a working hypothesis, to be confirmed or refuted case by case. If the word is used broadly, it means the antithesis of what many rural people and outsiders will agree in saying no to. It can include food, health, a strong family, wealth and income. It can be described as a level of wealth and of stocks and flows of food and cash which provide for physical and social wellbeing and security against impoverishment. To this may be added access to basic goods and services, but while these are important, for the poorest they may only come second to subsistence and security. 'A plate of basic needs does not whet the appetite while the prospect of a secure a plentiful livelihood according to the familiar pattern is an everyday motivation' (UNRISD, 1979, p. 10). Nor are livelihoods and basic needs everything. There is also the quality of living and experience – the value people set on the familiar, on being needed, on a purpose and role in life, on love, on religious observations, on dancing and song, festivals and ceremonies, on things in their seasons, and bringing in the harvest. Perhaps the most one can say is that for the full enjoyment of these, secure and decent livelihoods may be necessary but not sufficient on their own.

This emphasis on livelihood fits the evolution of ideas about 'rural development'. In outsiders' usage this means 'desirable change in rural areas'. What is considered desirable differs, by country, by region, by person, and over time. Rural development has been identified variously with economic growth, with modernisation, with increased agricultural production, with socialist forms of organisation, and with services for basic needs such as health, education, transport and water supply. As is very well known, the development thinking of outsiders has shifted from the view that growth and modernisation would be enough, with benefits trickling down to the poor, to the more realistic if depressing view that sometimes growth and modernisation make the poor poorer; that the main gain from increased agricultural production often goes to urban populations and the rural rich; and that the better off and more powerful benefit more from rural services than do the poor and weak.

In consequence, the sense in which 'rural development' is used has also changed. The World Bank Sector Paper on Rural Development, published in 1975, went this far:

Rural development is a strategy designed to improve the economic and social life of a specific group of people – the rural poor. It involves extending the benefits of development to the poorest among those who seek a livelihood in the rural areas. The group includes small-scale farmers, tenants and the landless.

(1975, p. 3)

The achievement of World Bank endorsement and propagation of this definition should not be underestimated. It goes further than many critics of the World Bank might have believed possible. At the same time, it reflects the thinking and power structure of the core. Women are not mentioned; nor are children. Strategies are designed, unless otherwise stated, by the powerful; and it is they too who *extend* the benefits of development *to* the poorest. This side of rural development is important, but it should be balanced by a reversal of power and initiative. A complementary definition might be:

Rural development is a strategy to enable a specific group of people, poor rural women and men, to gain for themselves and their children more of what they want and need. It involves helping the poorest among those who seek a livelihood in the rural areas to demand and control more of the benefits of development. The group includes small-scale farmers, tenants, and the landless.

The initiative starts with outsiders; but the aim is to transfer more and more power and control to the poor.

These definitions are at a general level. Discussion is often at a lower level of what are really sub-objectives. Some of these are increasing productivity, improving equity, and maintaining and enhancing the renewable resource base of the environment. But, these are means, not ends, and can be tested against the primary objective of secure and decent livelihoods. No sub-objective is automatically good. The productivity of land or labour can be enhanced in ways which impoverish: a farmer can mechanise his agriculture and increase production but destroy the livelihoods of labourers. Production that leads to net impoverishment is not development. Production is a means, not an end. The end is in the lives of people. Equity, too, deserves scrutiny. A dramatic levelling down, as in Kampuchea under Pol Pot, may reduce inequalities while destroying livelihoods; at the extreme, many are equal because many starve. Again, the environment exists for man, not man for the environment. If desperate poor people 'mine' steep forest lands or fragile savannahs, securing precarious livelihoods for a time at the cost of long-term loss of potential, one must be careful about inconsistent condemnation. Is it worse for those who would otherwise starve to mine hill slopes, or for those who are in no way threatened with starvation to mine non-renewable oil reserves? Where the survival of a poor household is at stake, who can condemn it for living at the cost of the future? What outsider, in the same circumstances, would not do the same?

In practice, though, outsiders who are removed from and do not perceive the lives of poor people often treat means as ends. Again and again agricultural production is seen by technocrats as an end in itself, regardless of whether poor people can grow the food themselves or buy it; equality is treated as an end in itself in some of the more doctrinaire socialist practice, even at the cost of food and income for the poor; and protecting the environment is implicitly held by devout conservationists to be a value higher than the livelihoods of the poor. Now there are, of course, national and urban interests in agricultural production and in the long-term productivity of land and water. But many outsiders are conditioned to give more weight to interests which are national and urban rather than local and rural, and to give least weight of all to the poorer rural people, trying to survive and to achieve some minimum decent standard of living. Deprivation for the rural poor is accepted in order to benefit urban and other interests. But priorities are suspect, to say the least, when they serve the interests of the powerful at the centre against those of the weak at the periphery. The priorities of the rural poor are not general – they are not agricultural production, equality, or the environment. They are particular, immediate, and personal. There are always trade-offs between multiple objectives. But really putting the poorer rural people first reorders outsiders' thinking so that what before were seen as ends are now seen as means – means to the over-arching objective of enabling the poorest to demand and control more of what they want and need.

Outsiders' interventions

Conventional logic would move from objectives to methods of analysis for the choice and design of action. But choice and design can be pre-empted by assumptions about the means and methods of implementation. The way people think about rural development is conditioned by the habits and experience of acting in certain ways. Many practitioners have a standard reflex to see action as carried out by bureaucracies. This 'programmatism', as it has been called by M. P. Moore (1976, p. 28) is deep and widespread. 'The problems of rural development' evoke responses in standard forms: plans, projects and programmes to be implemented by organisations, especially those of government. This is so obvious that it easily goes unrecognised.

Rural development *plans* are found for the regional or provincial, district or subdistrict and block levels. Typically, over the past two decades, the objectives of such plans have not been clearly stated. Although the development sought has often been described as 'integrated', the plan document which provides the blueprint is schizoid, with compilations of data coming first, and an unconnected shopping list of projects following. Rarely have such plans been implemented.

For implementation, *projects* and *programmes* have a better record. Projects are implemented by both government and non-government organisations in defined geographical areas, and are sometimes pilot or experimental, receiving special attention and evaluation. Programmes cover wider geographical areas and are usually implemented by government bureaucracies in such fields as health (vaccinations, new health posts, community health workers, sanitation, etc), and agriculture (the diffusion of new crop varieties, agricultural credit, artificial insemination, fertiliser trials, advice on new cultivation practices, etc); and examples could be added from domestic water supply, community development, irrigation, forestry and other departmental spheres.

Plans, projects and programmes are often nowadays intended to benefit 'the rural poor', 'the vulnerable groups', 'the backward classes'. Projects which are targetted to such groups, and especially those run by voluntary agencies, have had some successes. Programmes run by large-scale government field bureaucracies have a less good record. It is useful here to distinguish two types of programme. The first and most common type can be described as 'spread-and-take-up'. The imagery is of a service being pushed out from the centre and taken up by people further and further into the periphery. Schools, health centres, agricultural extension, roads, and community development have all followed this model. Those who take up and make use of the services are at first those who are better placed geographically, socially and economically. But the aim is that by pushing hard, and reaching down and out, all will at last be reached.

A second and less common approach has been to start from the other end, with programmes designed for the last first: the Botswana Government's programme for 'extra-rural' areas, so called because they are so remote; the Indian Government's Integrated Rural Development Programme; nutrition programmes designed to assist precisely those who are worst nourished; the *antyodya* programme in parts of India in which each village was asked to identify its poorest families and to do something to assist them; relief work in famines and following disasters – these are examples of programmes which are intended to put the last first.

Both spread-and-take-up and last-first programmes are vulnerable to interception by the elite. The Small Farmers Development Agency and Marginal Farmers and Landless Labourers programmes in India slid up the scale, benefiting people who were better off than the intended population (Pant, 1982). Loans meant only for the very poor are taken up by their patrons; food intended for the poor is appropriated and sold by the wealthy (BRAC, 1980). But negative social science has overstressed the failures. There is a positive side. In three recent decades average life expectancy in the developing world has increased by nearly 30 per cent, from 42 to 54 years, and the proportion of adults who are literate has risen from about 30 per cent to more than 50 per cent (World Bank, 1980, p. 1). These achievements are, of course, not only the result of administered programmes; but programmes for health, agriculture, and education have played a major part in them. 'Spread-and-take-up' programmes remain one of the most obvious and most powerful tools available to governments for alleviating rural poverty.

But programmatism has drawbacks. It tends to be standardised, top-down, authoritarian, and unable to adapt to local conditions. Because there was a national drive to grow maize, agricultural extension staff in Zambia tried to persuade farmers to plant it in all areas, even where it neither grew well nor paid; and there are many similar examples from other parts of the world. As Moore points out too, programmatism drives out consideration of other approaches. Two such omissions stand out: prices and terms of trade; and enforcing the law.

First, prices are perhaps the most powerful weapon for augmenting rural incomes.[2] They often make massive transfers of incomes from rural to urban populations: monopoly marketing of cash crops which allows surplus to be creamed off, to benefit the bureaucracy of the parasital[3] which manages the marketing, as notoriously with some West and East African commodity marketing boards; low regulated urban prices for rural produce (food crops, fish, charcoal, honey), but high unregulated rural prices for urban produce because of scarcities and costs and difficulties of transport as in Zambia (ILO, 1981); and an overvalued foreign exchange rate. All favour urban against rural populations. An overvalued foreign exchange benefits the urban middle classes by keeping down the prices of the imported goods they buy, and discriminates against the rural population by giving them less for the commodities they produce for export, such as jute, sisal, tea, coffee, cocoa, coconut oil, palm oil, rubber and cotton.

It is in the power of most governments to change prices, raising what is paid to rural producers. The political problems of alienating urban middle classes, students and workers through higher food prices are well-known, and the needs of the urban poor have to be weighed. But these considerations should not obscure the potential of this approach. Devaluation is another method: in some economies this can be a sharp and simple means of redistributing incomes from the urban middle classes to the rural areas. Devaluation also raises production costs (of imported fertilisers, mechanical equipment and fuel) of the larger and more prosperous farmers, but this may be more than offset for them by the higher prices they receive for export commodities. The gains of the rural poor come through less capital-intensive techniques in agriculture which sustain more livelihoods and raise wages and through secondary effects of the higher rural incomes of others. Those who are micro-producers of export commodities also benefit directly. In the practical repertoire of governments, however, devaluation and high prices for rural produce do not often receive the attention they deserve.

The second neglected mode of intervention is enforcement of the law. The potential for benefiting the poor here appears greater in South Asia than in Africa. Land reform in India provides on example. There are land reform laws regulating land ceilings, terms of tenancies, and the like, on the statute books of the Indian States; and while disagreeing on other points, almost all observers agree that effective land reform, more than any other single measure, would diminish rural poverty in India. Where a resolute attempt is made to implement the law, as in West Bengal, something is achieved. Elsewhere, land reform legislation may have discouraged larger landowners from buying more land and encouraged them to sell off land to tenants rather than risk transfer to them without compensation (Cain, 1981, pp. 447–8). Or, to take another example, interest rates in India are regulated by law, but the upper limits are rarely observed in rural areas. Legal aid to the poor resulting in successful cases against extortionate moneylenders, is a means of bringing down the going rate of interest. With both land reform and interest rates, one effect of threats of enforcement can, thus, be to benefit the poor.

But conventional analysis and programmatism tend to be limited to proposals which can be carried out by existing organisations. This points away from prices and the law. Prices inspectorates either do not exist in rural areas or are insignificant compared with departments of agriculture, health, or community development. Still less does one find rural legal departments; lawyers vie with town planners and sewage engineers in their urban bias. Without an organisation, programmatism is impotent. Thus prices and the law are overlooked although they are some of the most effective ways of transferring resources, control and initiative to the rural poor.

Analysis for action

Errors enter and chances are missed not only through habits of action, like programmatism, but also through defects in analysis. Ways of thinking about rural development are rarely thought about. Though basic, they are taken for granted. Yet there are many different ways of analysing rural environments and rural deprivation. Of these four stand out: costs and choices; causes and constraints; opportunities; and political feasibility.

i) Costs and choices

Questions of costs and choices are basic to practical thinking about action. This is obvious but often disregarded. Resources are scarce in relation to needs and opportunities. Costs and benefits have to be assessed and choices made. Yet again and again the cost side of decision-making is poorly analysed and the range of choices neglected. The costs considered are often only the obvious financial ones, but frequently more important though less well recognised are costs in staff time and administrative capacity. Especially with field-level staff there are opportunity costs – that is costs in terms of alternative benefits foregone – in their deployment on one programme rather than another. Choices may be made between alternative uses of funds, but rarely between alternative uses of staff, who are treated as infinitely elastic. This exercises yet another bias against the poor. Under pressure of impossible demands from above, field staff leave much undone. They choose not to do whatever will lead to least complaint if neglected, and this is almost always work with or services to the poorer who lack power and resources. Instead field staff concentrate their attention on those tasks which are least irksome, most congenial, privately most profitable, most easily inspected by superiors, and most likely to provoke sanctions if not performed; those tasks, in short, which involve the more accessible local elite. A failure to choose between incompatible uses of field staff time usually discriminates against those who are poorer.

The failure to think clearly about costs and choices is common when core people prescribe for the periphery. Two forms are prevalent.

The first is advocating everything at once. Two examples can suffice. Erik Eckholm, in *Losing Ground* (1976), describes what he calls 'the ultimate vicious circle' of rapid population growth, miserable social conditions, and environmental degradation and concludes that 'The only alternative at this stage of human history is to *simultaneously* meet this quandary *at every point* along its circumference, in an all-out effort to turn the negative chain reactions into positive ones' (1976, pp. 23–4, my emphases).

Similarly, J. K. Galbraith concludes *The Nature of Mass Poverty* (1979) with these words:

> ... Mass poverty ... is a tightly integrated phenomenon. And so, accordingly, is the remedial action. The breaking of accommodation and the provision of the several escapes – within the equilibrium and culture of poverty, to alternative urban employment, from the country – are parts of an organic whole. And the relevant help from the rich countries – for the requisite educational, agricultural and industrial investment – equally belongs. *No one can say that one part is more important than any other.* But, likewise, all must be on the agenda of the poor countries and the conscience of the affluent lands. *No remedy for poverty can be excluded from what is an organic whole.*
> (1979, p. 139, my emphases)

Both Eckholm and Galbraith are trapped by a high level of generalisation, a complex subject, and perhaps a commendable desire to be brief. Eckholm advocates simultaneous action on all fronts, for Galbraith no one can say that one part is more important than any other. Statements of this sort tend to crop up near the ends of chapters or books when authors fear they have left something out, or been too dogmatic, or both; and no one will want to throw stones for fear of broken glass at home. But they do sidestep the issues of choice; and they are unlikely to reflect the views of the authors when confronted with concrete situations. It is not the practice of (successful) generals to attack simultaneously on all fronts because they cannot say that one is more important than another.

The second failure of analysis occurs in the ritual call for integration and coordination, and even maximum integration and maximum coordination. These words slip glibly off the polished tongues of practised non-thinkers. They can be strung together in alternate sentences to give a semblance of solidity to a smokescreen of waffle. Maximum integration could mean all departments doing everything with all other departments at all levels; and maximum coordination could mean everyone meeting everyone else and discussing everything. Both integration and coordination have high costs. Both involve choices by default – choices not to use funds, administrative capacity, and staff time, in other ways. Both can blunt action and demoralise. There may be a law that the chances of a memorandum or report being implemented vary inversely with the frequency with which the words integration and coordination are used. For they evade the hard detailed choices of who should do what, when, and how, which are needed to make things happen.

ii) Causes and constraints

Two further ways of thinking are common among academics who are drawn towards practical analysis for rural development. These concern causes and constraints.

With the first, the implicit reasoning is that the causes of rural poverty can be identified and then attacked. This, it is sometimes said, may get to the roots of poverty. Getting at these roots, weakening or removing the causes, will then also weaken or remove rural poverty. A medical analogy and medical language are often used. There is first a problem of 'diagnosis' which is followed by 'prescription'. Implementation of the prescription (the equivalent of clinical treatment) will then, it is assumed, overcome a pathological condition and restore a condition of health. In India, the phrase 'malady-remedy analysis' has been used and reflects this pattern of thought.

The second way of thinking concerns constraints and obstacles. There are 'limiting factors' and 'obstacles to development' (UNRISD, 1979, p. 1). Analysis here, it is implied, will identify what is impeding development. If the constraints can be eased and the obstacles removed, then development will be free to take place.

These two ways of thinking have strengths and weaknesses. On the positive side, they exercise some discipline over analysis. It is a major step to ask seriously why rural poverty exists, what maintains it, and what impedes poorer people from becoming less poor. But there are weaknesses too. One danger is an unsubtle frontal approach. Crude population programmes are an example, where the pathological condition or obstacle is seen as too many children, and the cure or means of overcoming the obstacle is seen as a draconian family planning programme. Thus Paul Ehrlich in *The Population Bomb*:

A cancer is an uncontrolled multiplication of cells; the population explosion is an uncontrolled multiplication of people. Treating only the symptoms of cancer may make the victim more comfortable at first, but eventually he dies – often horribly. A similar fate awaits a world with a population explosion if only the symptoms are treated. We must shift our efforts from treatment of symptoms to the cutting out of the cancer. The operation will demand many apparently brutal and heartless decisions. The pain may be intense. But the disease is so far advanced that only with radical surgery does the patient have a chance of survival.
(1971 [1968], p. 104)

The implications of such an approach can be questioned not merely on grounds of compassion. They also have a directness which is insensitive and unlikely to work. As is now well accepted, the greater challenge with problems of population is to establish the preconditions for lower fertility – including lower death rates, better education, especially female education, better health facilities, and better livelihoods which will predispose people to wish in their own perceived interests to have fewer children, and thus to wish to avail themselves of family planning advice and facilities.

Another danger of frontal approaches is despondency. The diagnosis may be that the poorer people are locked into exploitative relationships which cannot in the foreseeable future be broken. The frontal solution is land reform in which the landless receive land, but there may be no early prospect of this.

The blockage or constraint is the rural and urban power structure. There is no immediate way in which that can be altered or overthrown, and the analyst gives way to cynicism and despair.

Yet another danger is focussing on one cause or blockage to the neglect of others which may be more critical. Professionals are conditioned by their training towards this error (see pp. 22–3 and 179–80). A doctor concentrates on primary health care, an agriculturalist on the spread of a new crop variety or technique of cultivation, a forester on protecting the forest, an engineer on road construction and maintenance, each in the name of development. Each encounters different obstacles. Each may readily assume that if those particular obstacles can be overcome, development will be achieved. This contrasts with the over-inclusiveness of Eckholm and Galbraith. But in fact, many influences interlock to impede development. Removing or weakening one may have little effect; it may even turn out to have been redundant.

A more insidious danger of thinking of causes and constraints is falling into the error of seeing poverty as a deviation from a natural path. Treatment of the disease then restores the natural condition of health; removal of the blockage allows the normal process of development to flow on. But poverty is not like that. Poverty is itself a natural condition, a result of physical conditions and human nature, including acquisitiveness for self, family and group. The problem is much, much deeper than that of a sick patient or a blocked stream; and far more imagination, ingenuity and will are needed to overcome it.

Despite these dangers, analysis of causes and constraints has its value. This lies especially in seeing how cost-effective it will be to weaken or eliminate one cause or constraint rather than another, or how several are linked and should be attacked together or in sequence. It can also provide tests of feasibility for action proposals through identifying causes or constraints which will impede, neutralise, or divert benefits intended for the poorer rural people.

iii) Finding and making opportunities

Much analysis is problem-oriented. The danger then is that it reinforces negative social science. Identifying constraints, maladies, or causes, or examining the problems of political will, are liable to be depressing. They are likely to identify why nothing will work unless there is an (improbable) change in the system as a whole. Perhaps their negative trend is a warning, too, against the abstract and academic. Faced with concrete situations, outsiders are more likely to look not for problems but for potentials, not for obstacles but for opportunities.

There is, it is true, a sense in which problems can be opportunities. The Chinese expression for crisis has two characters – described by Mark Svendsen as one meaning problem, and the other opportunity. Crisis brings forth special efforts – the 'creativity' which Hirschman has argued is as habitually underestimated as are the difficulties it is mustered to overcome (Hirschman 1967). Famine relief is turned into food for work to build infrastructure to help avert future famines; desert creep provokes afforestation which provides livelihoods and fuel; a flood triggers proposals for a dam and irrigation. But many problems are better not 'solved': bad projects which produce little and drain national revenue; a refusal of farmers to plant an uneconomic crop; attempting to market a product for which there is little demand; trying to make an organisation work when the costs are high and the chances of success low. The costs and benefits of solving problems have to be weighed against the costs and benefits of alternative actions.

Two relevant ways of thinking are suggested by the geneticist C. H. Waddington in his book *Tools for Thought*. Both are concerned with system and process.

The first is described by Waddington as 'soft spots' (an alternative more popular image is 'weak links in a chain'). Soft spots are points in networks where alterations are likely to have more profound effects on a system than others. Waddington wrote:

> If one is trying to alter a system which has some inbuilt buffering, one of the most important first steps is to try to locate these 'soft spots'. There has been a good deal of theoretical discussion about how to locate them, or preferably how to measure the sensitivity of each particular link in the network. The most important result to emerge is that the sensitivity of a particular link is not a fixed characteristic of it, but depends on the state of the rest of the network.
> (1977, p. 91)

Applied to attempts to reduce rural poverty, this implies careful analysis case by case. It also implies, notwithstanding the primacy of livelihoods, that the points of intervention in trying to change the web of deprivation (see pp. 111–14) will by no means always be the same. A cost-effective intervention will be based on an analysis of the network and a search for soft spots.

The second way of thinking is in terms of what Waddington calls an epigenetic landscape. This is a landscape in which there are valleys which branch as they descend. Systems (nations, regions, communities, households, for our purposes) can be seen as moving down such valleys.

> Sometimes one knows that there is a branch point ahead of the system, and that if one can give the system a push at the right time, it can be diverted into one or other of the alternatives in front of it. The point to notice here is that it is in general no use giving the push too early: if you do, the system will have got back to the middle of the valley again before it reaches the fork, and the effect of the push will have been dissipated. The period just before the branching point, during which a push will be effective in diverting the system into one or other path, is known in biology as the period of 'competence'. It is no use trying to act on the system to divert it into a particular branch until it has become competent to respond, by going down the valley towards which you have pushed it. Equally, of course, it is not advisable to leave the push until too late. Once the system has started to go down one of the branch valleys, if you still want to divert it into the other you have to push it right over the hill between them. Effective revolutionaries, like Lenin, have been brilliant in choosing just the right time to give a push to a society coming up to a branch point in its stability system.
>
> (*ibid*, pp. 110–11)

In rural development, this suggests attention to timing and irreversibility. The trick is so to analyse the processes of change as to see what branches a system is reaching and what pushes can be applied at what time so that the valleys entered benefit the poorer people. This sort of thinking is not common. It applies to ecological change which is irreversible or difficult to reverse: desertification in the Sahel, cutting down tropical forests in the Amazon basin, removing the forest cover needed for cocoa growing in Ghana to grow food, thus preventing reversion to cocoa later. These are 'valleys' from which escape becomes increasingly difficult; and the fact that poor people are so often driven into them through sheer necessity to survive makes it even more vital to catch and control such processes early on.

The identification of critical periods of competence applies with great force where new resources, new technologies and new activities are being introduced and which, once appropriated by elites, will be very difficult to detach from them. Perhaps the greatest opportunity India has for establishing landless families on land is at the time when new canal irrigation is introduced; the productivity and value of land rise sharply, and the landholdings of larger farmers can be reduced in size without their losing in absolute terms. India proposes to double the area commanded by canal irrigation by the end of the century, adding some 30 million hectares. Even allowing for shortfalls in implementation, this programme has a potential for settling millions of poor families. But to do this, concerted action is required at the right moment, before and during the arrival of the first irrigation water. Similarly, in Bangladesh, as low lift irrigation pumps which draw on communal water in canals and drains are introduced, they may be appropriated by those who are better off, or they can be allocated, as some are, exclusively to groups of the landless. Or again, in India, social forestry and community forestry programmes usually take two forms: either trees are planted on private land, in which case the owners (usually the better-off farmers) benefit; or they are planted on common land but without clear definition of rights. Who will benefit and how is left for the future, and the rights of weaker families are not established. The opportunity with the common land here is to recognise and register from the start the rights of poor families to exclusive, preferential or at least equal use and benefit from the trees. Much the same applies with other commons, including groundwater. In many rural environments where there are still unappropriated resources, the critical period when they can be allocated to the poorer people without a major upheaval often passes unnoticed and fast.

A search for opportunities can generate an agenda for action. Many different lists could be drawn up. Many approaches – from health, education, communications, water supply and so on – are possible. Those which follow are limited to resource-based opportunities for generating or strengthening livelihoods at the local level, and are meant to illustrate not summarise:

– exploiting common property resources: setting aside forests, woodlands, stands of bamboo, reedbeds, areas that flood, waste land, communal land, hunting, gathering and fishing rights, surface water, groundwater, etc., for the poorer people, ensuring that the means and rights to exploit these are reserved to them;

– releasing time and energy by reducing drudgery: through improved tools and/or processes, cutting down the time and energy required for drawing and fetching water, collecting and carrying firewood, cultivation, processing food, and cooking, so that more land can be cultivated, or land can be cultivated more intensively, or other work can be done;

- seasonal support to create livelihoods: by enabling poor households to work and earn or otherwise be productive in slack seasons, lifting them out of a seasonal trough, and over a threshold of stocks and flows of food and income, to achieve a secure and decent livelihood;

- micro-capital for groups or households: providing, either free, subsidised or on credit, items of capital to poor households – small-stock for rearing, animals for transport or draught, ploughs, tools, handpumps for irrigation by households or larger pumps for irrigation by groups, hand or foot-operated machines for agricultural operations especially threshing and other crop processing, etc;

- irrigation: to reduce risks, raise yields, grow crops at shorter intervals, reduce migration to the towns, increase employment and raise wages;

- water-harvesting: improvements to micro-catchments through small-scale storage, contour ploughing and ridging, mixed cropping, grassed waterways, and other means of improving moisture retention and infiltration, thus reducing risks, and increasing yields;

- crop varieties: breeding and disseminating varieties which fit farming systems, are disease-resistant, drought-tolerant, high-yielding and low risk.

The list could be made much longer. The point, however, is made. There is a vast positive side to the problems of rural poverty, the side revealed by searching for potential and exploiting opportunity, a side visible to practitioners and natural scientists but often overlooked in negative social science.

iv) Political feasibility

The gravest neglect in analysis for practical rural development has been political feasibility. Again and again, projects and programmes have been designed and targetted for the poorer people, only for the bulk of the benefit to be captured by the less poor. In the already quoted words of Christ, according to St Matthew, 'Unto every one that hath shall be given, and he shall have abundance; but from him that hath not shall be taken away even that which he hath' (The Bible, St Matthew, xxv, 29).

This tendency has been described by Andrew Pearse (1980) as the talents effect[4], and by Ward Morehouse (1981) as the refraction effect. As this reality became more evident during the 1970s, it became routine to gloss it in terms of 'lack of political will'. But 'lack of political will' usually means that the rich and powerful failed to act against their interests. 'Political will' is a way of averting the eyes from ugly facts. It is a conveniently black black box. Just as 'integration' and 'coordination' stop short of asking who does what, when, where and how, so 'political will' stops short of asking who gains and who loses what, when, where and how. But if this question is not asked and answered, projects and programmes will continue to be intercepted, distorted and captured by powerful interests and local elites. Political will, in the sense of these forces, is a brick wall. When a project or programme bangs its head against the brick wall and the wall does not fall down, one response is to bang harder. Such frontal attacks usually fail and feed either self-deception or pessimism. Another approach is to look at the wall in detail, to search for and dislodge loose bricks, to seek ways round the side, to judge whether it can be jumped over, or sapped from below. Often, in the short term, the most effective approaches are not total and frontal, but piecemeal and oblique.

For such approaches, it helps to identify who will gain and who will lose. A simple table can indicate different ways in which benefits may be distributed between two categories of people: the rural elite; and the poorer rural people. Some outcomes are shown in Table 6.1. Type A programmes reinforce the control and augment the wealth of those who are already powerful and wealthy, at the cost of the poor. In type B programmes, all gain. In type C, the rural elite neither gain nor lose, but the poor gain. In type D, the rural elite lose and the poor gain. In the absence of strong political organisation by or on behalf of the poor, there is a gradient from high feasibility with A, down through B and C to very low feasibility for D.

To be realistic, refinements must be added. Situations are dynamic. Gains and losses cannot be seen simply in terms of whether people are better or worse off at time $T + 1$ than they were at time T. People themselves see their gains or losses not only like this but also in terms of what might have been. Thus if a new resource becomes available and is allocated to the poorer with no gain to the rural elite (type C), the rural elite may object and interfere not because they will be worse off in absolute terms, but because they will lose the benefit they might have had. Similarly, with some but not all type B programmes, where all gain, the rural elite may object and interfere because the benefit going to the poorer would, in other circumstances, go to them. They therefore see the situation as a zero sum game, although they are already benefiting. They are likely to object even more if through their benefits the poor become more organised,

Table 6.1 Benefits distributed between the rural elite and the poorer rural people

Type	Rural elite	Poorer rural people	Examples
A	Gain	Lose	Allowing or enabling elite to appropriate common property resources (land, ground-water, fish, forests, pasture, reeds, silt, stone, etc.); denying these to others.
			Technology with net livelihood-displacing effects (Modern Rice Mills, combine harvesters, herbicides, tractors, etc., depending on local conditions)
B	Gain	Gain	New services accessible to all (health, water, education, shops with basic goods, etc.).
			Most public works which create new infrastructure
			New irrigation or other technology which increases employment and raises wage rates.
			Canal irrigation reform from which all farmers gain (see p. 188).
			Development of a common resource where all share in its exploitation.
C	No change	Gain	Extending the coverage of spread-and-take-up programmes to be accessible to more of the poor.
			Credit for enterprises for the poor.
			Subsidised rations for the poor (as in Sri Lanka).
			Appropriate technology (seed varieties, tools, cropping systems, etc.) for resource-poor farmers.
D	Lose	Gain	Land reform with inadequate compensation.
			Implementation of minimum wage legislation.

articulate, and politically active in demanding their share and their rights.

Even if a search is made for type B and type C programmes, there will still be conflicts of interest. Besides the rural elite and the poorer rural people, there are other interest groups including local politicians, local bureaucrats, central politicians, and central bureaucrats. Judith Heyer (1981, p. 215) has argued for an analysis of interest group conflicts and taking these into account through a process of bargaining, compromise and explicit concession to groups that normally lose out. Nor are broad categories of interest groups adequate. Charles Elliott (1982) has shown for the Dal Lake in Kashmir how conflicts of interest lay between nine groups – reed cutters, fishermen, gardeners, houseboat owners and operators, hoteliers, property developers and owners, traders, farmers, and the Forestry Department. These conflicts of interest included some between poor groups themselves – fishermen who wanted more fish versus reed cutters and gardeners whose activities reduced the fish population. The distribution of gains and losses differed according to the intervention proposed, and only one of the five interventions analysed – the introduction of grass carp – involved no loss to any group. In this example, as elsewhere, who stands to gain and who to lose affects political feasibility.

Political feasibility is not, however, part of standard programme or project appraisal. There are well established procedures for assessing technical, financial and economic feasibility, deficient though their use often is, but none in any formal sense for assessing the power and interests of groups, how they converge or conflict, and how they will support or impede the achievement of a project's objectives. Group interests and conflicts are, however, as much a part of the environment as are climate, soils, water, cropping systems and the like. The ignoring of the power and interests of local elites, more perhaps than any other factor, has been responsible for failures to benefit the poor. The opportunity now is to use the insights of negative social science positively. An analysis of interests generates a new set of questions and a new repertoire of tactics involving trade-offs and compromises as part of a new realism. Secure and decent livelihoods for all remain the long term objective, and for that type D solutions may eventually be necessary. But in the shorter term, the aim may be rather through B and C type programmes to gain bridgeheads and salients here and there, strengthening the hand of the poor, and enabling them to act more for themselves.

Power and the poor

Many outsiders prefer diagnoses and prescriptions which gratify them. The most immediately gratifying are those where direct action yields quick results against visible physical weakness. Urban elites give enthusiastic support to eye camps (where those who are blind receive sight) and to feeding programmes for children (where the malnourished are fed). Rotary Clubs, Lions Clubs, and their equivalents, subscribe to such campaigns where direct physical benefits can be seen. Similarly, Albert Schweitzer, Baba Amte, and others are well known for their sacrifice and devotion to direct action against the conspicuous horror of leprosy. And Christian and other medical missions around the world have set out to cure the sick. Nothing that follows should undervalue what they and others like them have done and will continue to do. But they attack symptoms, not causes. They do good work. But at the same time, direct action against physical weakness may distract attention from less palatable, more difficult, and in the longer term more effective measures.

Outsiders also prefer diagnoses and prescriptions from which they and those like them will gain and not lose. Direct approaches to the five dimensions of rural deprivation vary in

Table 6.2   Acceptability of rural development approaches to local and other elites

Dimension to rural deprivation	Examples of direct approaches	Acceptability to local and other elites
Physical weakness	Eye camps Feeding programmes Family planning Curative health services	High
Isolation	Roads Education Extension	
Vulnerability	Seasonal public works Seasonal credit Crop insurance Preventive health	
Poverty	Distribution of new assets Redistribution of old assets	
Powerlessness	Legal aid Enforcement of liberal laws Trade unions Political mobilisation Non-violent political change Violent political change	Low

the degree to which elites, especially local elites, gain or lose from them, and thus in the degree to which they are acceptable.

The best mixes and sequences of feasible actions vary both geographically and by stage of development. Attacks on physical weakness, isolation and vulnerability sometimes deserve priority, especially in the early stages of rural development. But later, and where there is widespread deprivation, direct attacks on poverty and powerlessness may be essential. Ways may be sought round, under, or over the brick wall of political will and entrenched interests. Often, though, there will be no effective alternative to enabling the poor to mobilise and organise, to demand, gain and maintain control over assets and income.

Altruism, idealism, and a spirit of service animate some outsiders, and some members of local elites; but it is too easy to perceive such people selectively. They are, after all, just the sorts of people met by rural development tourists; and they are a minority. Optimism generated by the achievements of the few must be tempered by the cold reality of the entrenched power of the many and of crude self-interest, whether individual, family, or along the lines of race, nationality, class, caste, religion, ethnic group, or gender. The blinding of police prisoners in Bhagalpur District in India in 1980 was shocking in itself, but it was more shocking that so many evidently were not shocked (Niesewand, 1980). There are many who wish to keep the poor weak, ignorant, vulnerable, poor and powerless. M. N. Srinivas records that he has heard

> ... a powerful headman dismissing the idea of building a new school building on the ground that it would only teach the poor to be arrogant. The same man wanted electric power instead for the village as it would enable an industry to be started. And he would make sure that he would start the industry.
> (1975)

Srinivas' headman may have judged well in his own interests: education may indeed be a key point of entry in setting off sequences of change which benefit the poorer. But many of the reversals needed will only come about with more than education: they require changes in the control of assets and in the distribution of power. It is a question of sequences and directions. The example of India is instructive, besides involving the largest number of poor rural people in any country. Indian rural society is much criticised for its exploitative structure. But in contrast with, say, South Africa, Indian law is liberal and points towards greater equity. In India and elsewhere, bridgeheads of organisation and resistance to exploitation have been established. Many voluntary agencies and some of those who work in government devote their energies to helping the poor to organise, defend themselves, and secure a better life. The Chipko movement in Uttarkhand in India has demonstrated how poor people can organise to resist the destruction of their livelihoods and environment, mobilising to join hands round their trees to prevent contractors cutting them down (Mishra and Tripathi, 1978). The land reforms in Kerala, Sri Lanka and West Bengal, for all their defects, are achievements from which poor people have benefited; and which have been carried out because of a popular power base for the government and a cadre of government staff which included people committed to the changes. Conversely, without an organised power base, and without outsiders' support, the rural poor remain vulnerable; and this means, in the economic and political conditions of the 1980s, that many will slide into even deeper deprivation.

There are many stages, sequences, gradations and subleties of change. There are communities where the rural elite will willingly identify those who are poorer so that they can gain specific benefits and where the elite are proud not to be poor and not to need the attention themselves. There are egalitarian communities with traditions of sharing (see e.g. Sahlins, 1974). But as these traditions weaken, and where they have already disappeared, then organisation, discipline, resistance and negotiation by the poorer people is often a precondition for social justice. The history of the trade union movements around the world is evidence enough of conditions where organised pressure has been needed. Most of the rural poor in most of the less developed countries do not yet have trade unions or indeed any organisation. Urban workers are organised, but rural are not. In Zambia, the mineworkers have a powerful trade union and can apply pressure for cheap food and low prices for rural goods; the rural poor – female-headed households, fisher people, charcoal burners, subsistence farmers, beekeepers – are almost entirely unorganised, being scattered, illiterate, poor, and preoccupied with scraping together some sort of livelihood. It is difficult or impossible for them to organise to offset urban interests. Within rural areas, too, it is often dangerous for labourers to hold out against employers for higher wages, or even for a statutory minimum wage; and where they do, as in parts of Bihar, death at the hands of landlords or police may be their reward. In such conditions, the struggle against poverty is, has to be, a struggle for political and physical power.

Non-rural outsiders can do many things. Representatives, allies and spokesmen for the poor can help. There is a long history of those, themselves neither poor nor weak, who stand up for those who are. Analysis by them, and with the poor, of the nature and extent of deprivation, of the forces which sustain it, and of the opportunities for attacking it, can sharpen strategies for intervention. But analysis and strategies are the easier part. More difficult are deeper changes – in values and behaviour.

Notes

1   For a useful typology on these lines, see Esman 1978: 9–11. He has pastoral societies as a fourth category.
2   See e.g. Elliott 1975, esp. pp. 356–60 and 364–66, Gore 1978, Lipton 1977 *passim*, and Maimbo and Fry 1971. Since this subject is extensively examined in the literature the treatment here is summary.
3   Perhaps a ration of one new word per book can be allowed. This word is not, so far as I know, in any dictionary; nor is it ever likely to appear in one. But I cannot resist using it, because it fits so many of the organisations which impede development.
4   The reference here is to the parable of the talents (Roman coins) in the chapter of St Matthew from which the quotation is taken.

CHAPTER SEVEN

The new professionalism: putting the last first

'You are old, Father William', the young man said,
'And your hair has become very white
And yet you incessantly stand on your head –
Do you think, at your age, it is right?'
Lewis Carroll, *Alice in Wonderland*, Ch. 5

Everyone is ignorant, only on different subjects.
Will Rogers, *The Illiterate Digest*

... and the last shall be first.
The Bible, St Matthew, Ch. 19, verse 30

For the rural poor to lose less and gain more requires reversals: spatial reversals in where professionals live and work, and in decentralisation of resources and discretion; reversals in professional values and preferences, from a 'first' to a last 'list'; and reversals in specialisation, enabling the identification and exploitation by and for the poor of gaps – under-recognised resources, and opportunities often lying between disciplines, professions and departments. Reversals require professionals who are explorers and multidisciplinarians, those who ask, again and again, who will benefit and who will lose from their choices and actions. New professionals who put the last first already exist; the hard question is how they can multiply.

Reversals

A theme of reversals runs through this book. For those who are poor, physically weak, isolated, vulnerable, and powerless to lose less and gain more requires that processes which deprive them and which maintain their deprivation be slowed, halted and turned back. These reversals have many dimensions. They include, for example, pricing policies, and rural-urban terms of trade (pp. 150–151). But three dimensions deserve special attention because they combine potential impact with feasibility. They concern reversals in space, in professional values, and in specialisation.

Spatial reversals

Reversals in space concern the present concentration of skills, wealth and power in the cores, draining and depriving the peripheries. They have two main complementary aspects: where people live and work, and seek to live and work; and where authority and resources are located.

We noted in Chapter One the inward flows of educated and experienced people along the gradients from peripheries to cores. At each point along these gradients people seek qualifications and opportunities for the next inward step: rural parents educate their children hoping they will gain urban employment; officials in districts seek postings to regional headquarters, those in regional headquarters try to get to capital cities, and those in capital cities try to join the brain drain to richer countries. The resulting movements are often anti-developmental. Poor parents in labour-scarce rural households give their children an urban-biased education and point them towards the towns where they exacerbate urban problems. Poor countries train doctors and engineers they need who then leave for countries where they are less needed. Cruel absurdities result: urban unemployment and misery while rural areas are short of labour; and poor countries deprived of the professional competence which they have trained at great expense.

On top of this, populations grow fast. Even in countries with a substantial urban sector, the rural areas will have to support much larger populations in future. The balance of misery, between deprivation in the town and deprivation in the countryside, may shift in the next decade, with slowing rates of rural-urban migration, and even perhaps some net flows back to rural areas. Such flows have in the past resulted from urban distress – economic decline, or revolution, or both. Examples are the migration out of Kampala under Idi Amin and the forced move of over a million people from Pnom Penh under Pol Pot. But for the future a more positive view can be taken. It is usually cheaper to generate rural than urban livelihoods, and economically more productive (Lipton, 1977). Whether people are treated as an end in themselves or as an economic resource, the pointers are in the same direction: towards priority for more and better rural livelihoods to support larger populations and to provide rural solutions to urban problems.

Decentralisation is one key to these spatial reversals. Many forces centralise power, professionals and resources in the urban cores: this is encouraged by national, urban and class interests; communications; markets and facilities; distrust of the peripheries and of those lower in the political and administrative hierarchies; personal interests in convenience, services and promotion (pp. 7–10); and the sheer weight of political and administrative influence. New investments, buildings, industries, and even agricultural processing are sited centrally, and scale, capital-intensity, and high cost correspond with the size and importance of the core where they are placed. Rural goods, taxes and people are sucked inwards, drained from the rural periphery more strongly and consistently than the opposite flow of goods and services is pushed or drawn outwards. If the economy falters, or goods are scarce, it is the periphery and the poorer people at the periphery who go short and pay more.

But these processes, though strongly determined, are not unalterable. With varying success, programmes of decentralisation have been implemented. Tanzania under the leadership of Julius Nyerere sent staff out from the capital, depopulating the ministries' headquarters, and allocated regional budgets with some local discretion in how they were spent. Decentralisation in Egypt and Sudan may provide further examples of what can be done when governments are prepared to devolve financial discretion. With strong leadership or strong local demands, it is possible, though difficult, to force funds outwards, to give more local discretion, to decentralise agricultural processing and small-scale non-agricultural production, to disperse, in short, parts of the cores towards the peripheries.

Where people decide to live and work, and where resources and discretion are located, depend on a host of individual decisions. Professionals, at many levels of hierarchy, from the lowliest field staff to presidents and prime ministers, daily make decisions which affect the spatial dimension. The extension agent may decide to take a bus to district headquarters to draw personal allowances or to bicycle to a remote village where his or her services are needed; a president or prime minister may decide to visit an industrial complex and receive the obeisances of captains of industry, or to go to a remote and poor rural area and listen to those who are poorest and most deprived. The extension agent may deliver all the subsidised fertiliser to nearby farmers, or may distribute it more evenly to include those who are further away. The prime minister or president may order flyovers for the capital city, or roads to reach remote villages. These are real choices. Professionals at all levels do have influence over choices such as these; and how they use that influence is affected in turn by their values and preferences.

Reversals of professional values

The values and preferences of professionals are, then, a point of entry. We have seen how they respond to the pulls of central location, convenience, opportunities for promotion, money, and power. But this is not all. They are also influenced by professional education and training.

Professionals are achievers. Were they not, they could not have got through school, training institute, college, or university. To succeed they learn to be sensitive to signals of approval or disapproval. They strive for recognition. Good marks and further education reward the accurate and faithful adoption and reproduction of the views and values of their instructors. So pupils, students and trainees keenly internalise the values of their teachers. To be sure, at their rare best, education and training encourage independent thought, disagreement with lecturers, and the choice of unconventional subjects for study and research. But usually the pressure is to conform – from students' families who have made sacrifices for the students' education, and from routinised and authoritarian teaching. Some universities resemble old-fashioned factories turning out a standard, third-rate, out-of-date, product – people with skills but no originality, with an ability to remember and repeat rather than to reflect and create, parrots more than true professionals. The PhD, with subservience to supervisor and slavery to method, can stunt and deform intellectual development. The damage done by university education is easily underestimated, not least because those whose opinions matter are initiates themselves, and have a vested interest in the system. But damaging or no, one major impact of it is the inculcation and embedding of professional values.

After university, those who pass from the academic to the practical culture find many of these values not just unquestioned, but sustained and reinforced by others, now in senior positions, who were processed earlier. For those who go from university into private practice, their senior partners, and the professional associations which they aspire to join, have clear expectations. For those who go into government, one hierarchical environment replaces another, as they pass from Faculty of Agriculture to Department of Agriculture, from Faculty of Veterinary Sciences to Department of Veterinary Services, from Faculty of Engineering to Public Works Department or Department of Irrigation, from Faculty of Medicine to Ministry of Health, from Department of Economics to Ministry of Finance or Planning.

Whatever their other differences the academic and practical cultures share the values of the rich and powerful cores. These are polar opposites of the values of the poor and weak peripheries. Table 7.1 lists some of these polar values and preferences. The reader may wish to add to or subtract from this list, and to rate professions, technologies, research, development programmes, and the preferences and behaviour of individual professionals, according to their location, in the two columns. It is also revealing to list pairs of value-loaded words which express and reinforce professional preferences. Two examples can be given.

First, the two lists might have been labelled 'sophisticated' and 'primitive'. But nowhere outside this paragraph are these words used in this chapter. They are not neutral, and both have done much harm. In its current usage, 'sophisticated' indicates approval, and refers especially to 'high' technology which is complex, capital-intensive, modern and so on, combinations of 'first' characteristics. But this is a recent meaning. The transitive verb 'to sophisticate', according to the Shorter Oxford Dictionary (1955 edition) means:

> To mix (commodities) with some foreign or inferior substance; to adulterate.... To deal with in some artificial way.... To render artificial; to convert *into* something artificial.... To corrupt or spoil by admixture of some baser principle or quality; to render less genuine or honest.... To corrupt, pervert, mislead (a person, the understanding, etc.).... To falsify by misstatement or by unauthorized alteration.

Language has played a trick on us, accommodating and affirming the cultural imperialism of the professional values of the 'first' world. Foreign 'substances' have become superior: what might earlier have been seen as adulterated, artificial and spoiled has become advanced, modern and good. Similarly, the usage of 'primitive' has shifted from 'original', and 'ancient', towards the negative sense of 'backward'. The new contrast of sophisticated (=good) with primitive (=bad) both reflects and reinforces biases against the 'last'. Elites aspire to the sophistication of the cores, and abhor the primitiveness of the periphery.

Table 7.1   Professional values and preferences

A For technology, research and projects

First	Last
Urban	Rural
Industrial	Agricultural
High cost	Low cost
Capital-using	Labour-using
Mechanical	Animal or Human
Inorganic	Organic
Complex	Simple
Large	Small
Modern	Traditional
Exotic	Indigenous
Marketed	Subsistence
Quantified	Unquantified
Geometrical	Irregular
Visible and seen	Invisible or unseen
Tidy	Untidy
Predictable	Unpredictable
Hard	Soft
Clean	Dirty
Odourless	Smelly

B For Contacts and Clients

High status	Low status
Rich	Poor
Influential	Powerless
Educated	Illiterate
Male	Female
Adult	Child
Light-skinned	Dark skinned

C For place and time

Urban	Rural
Indoors	Outdoors
Office, laboratory	Field
Accessible	Remote
Day	Night
Dry season	Wet season

The second example of value-loaded words is the use in India of 'major' and 'minor' in irrigation and forestry. Major irrigation refers to canal systems and commands of over 10 000 hectares;

medium irrigation is between 2000 and 10 000 hectares; and everything under 2000 hectares is minor. Major irrigation receives the bulk of attention; it is prestigious, involving the investment of large sums of money, the design and construction of large dams and canals, and the gratification of impressive and highly visible achievement. Minor irrigation receives much less attention, involving only small and scattered physical works, investment in smaller projects, and less visible and impressive achievements. Yet in 1981 minor irrigation accounted for some 32 million hectares, more than the 28 million hectares of major and medium irrigation. Minor irrigation was also growing faster, and cost much less per hectare.

In Indian forestry, timber resources have been classified as 'major forestry produce' and non-timber as 'minor forestry produce'. Sharad Sarin has pointed out that this distinction of 'major' and 'minor' has led to a certain orientation and perceptions

> ... which are far removed from reality. For instance, the entire orientation and thus organizational/administrative arrangements in the form of structures, systems, training of manpower, procedures, planning, etc. of the forest department, appears to have been around major forestry produce. Little attention has been paid to the management of minor forestry produce.
> (Sarin, 1981, p. 398)

Yet most of the 40 million odd tribals in India, among the poorest and most despised of people, directly depend on the collection of minor forestry produce, and for many of them it is a key source of livelihood. Moreover, its recorded value during the first half of the 1970s, even at the low prices paid for it, was over one-quarter of the value of total forest produce (ibid., p. 410). But 'major' timber involves larger lumps of money, the use of machinery, commercial contracts, and people who are well-off, organised and influential, while 'minor' forestry produce (leaves, seeds, gum, honey and the like) involves smaller lumps of money, collection by hand, informal sale, and those who are poor, unorganised and uninfluential. Not surprisingly, in both irrigation and forestry, the use of these two words – major and minor – diverts attention and resources away from 'last' things and towards those which are 'first'.

As in these examples, many of the 'first' preferences and values overlap and support each other. Many are so deeply part of the way professionals see things and work that they pass unnoticed. Take for example the male–female dimension. Male predominance and dominance in organisations is so marked and so widespread that to many men it is, quite simply, natural, and the question of deliberate action to increase the status and numbers of women staff does not arise. In most universities, research institutes and government departments, the great majority of staff are men, especially where 'first' subjects are concerned. Those few departments where women are numerous and sometimes predominant deal, in contrast, with 'last' subjects – nutrition, home or domestic science, childcare, handicrafts, and women themselves. Such departments have a low status. Some are no more than token gestures. A women's wing in agricultural extension is understaffed and underfunded and consequently ineffective. In an agricultural university, a home science department is a poor relation, staffed mainly by women and involved with 'last' subjects. The important, male, visitor to the university is taken to the higher status male-dominated departments while his wife is sent to be entertained by the female home science staff; these staff thus never meet, or are met by him.

University syllabi and research, and knowledge itself, are dominated by the 'first' column. A special reverence is reserved for quantification, and preferences shown for the quantifiable. Not surprisingly, much more is known about 'first' subjects than 'last' ones. Compare knowledge of space rocket technology with the ignorance of how female-headed households in remote rural areas contrive to survive tropical wet seasons. There has been an explosion of journals in the hard sciences; there is no journal of rural poverty.[1] More is known about computers than goat droppings: a large imported computer scores straight 'first', a local goat's droppings straight 'last'.[2]

At this point, misunderstanding might arise. I am *not* saying that third world professionals should abandon the 'first' list. I am *not* saying that poor countries should abstain from developing the expertise to manage 'first' hardware and to negotiate with foreign organisations peddling 'first' technology. I am *not* presenting a neo-colonial argument designed to perpetuate dependence. Nor am I advocating a naive Luddism. Complex, capital-using technology has had, and will continue to have, important applications in attacking rural deprivation. It took the electron microscope, scoring heavily with 'firsts', to discover the miniscule rotavirus (scoring heavily on 'lasts') which causes so much (rural, indigenous, organic, untidy, soft, dirty, smelly . . .) diarrhoea, accounting for perhaps one-third of the attacks on vulnerable weaning group of children. It took a large imported computer to reform Kenya's examination system, making it fairer, and providing information about performance which enabled backward districts and schools to do better (Somerset 1982). Other examples where contributions have already been made, or are promised, include remote sensing from satellites to identify good sites to bore for water (Tanzania), radio communications for the management of large irrigation systems to improve responsiveness to farmers' needs (Philippines), aerial photography for land consolidation (Kenya), refrigerated seed banks to preserve disappearing genetic resources (in several countries), and laboratory analyses to pick up trace element deficiencies in soils (in many countries). And other 'first' technology, such as lasar techniques for precision land-levelling in irrigated deltas, photovoltaic conversion of solar energy to pump water, and the like, should not be rejected automatically but carefully weighed on their merits. So-called high technology does have selective applications in rural development. I am not advocating a universal shift from 'first' to 'last'.

But in almost any field of professional concern, the biases are loaded against the attributes and things that are directly important to poor rural people. Endless illustrations are possible: colonial and post-colonial prejudices against small native cattle, small stock, shifting cultivation and intercropping; the long neglect of the diarrhoeas; cold water fishing research devoted to exotic trout for the recreation of elite fishermen rather than local fish for livelihoods for low status fisherpeople; forestry research on commercial teak for the few rather than fuelwood species for the many; or the continuing neglect of some subsistence crops. Often the automatic, unstated, and unquestioned assumptions are striking and shocking once they are recognised.

Cassava (manioc, tapioca) is the major staple of tens of millions of people, and the food of last resort of many more, and its processing is a laborious task for millions of women. Yet an FAO publication – Cassava Processing (Grace, 1977) – is devoted only to commercial processing; it is as though to the author the word 'processing' could only refer to 'first' processing by machines, in quantity, for the market. Yet such biases are not inevitable. Another FAO publication (French, 1970) is a sensitive, well-informed labour of love for a despised and rejected 'last' animal, even if its title – Observations on the Goat – is half apologetic.

It is a commonplace how the pursuit of 'first' values and technology favour the rich and the less poor. The social science literature on the green revolution,[3] even after discounting for negative social science, shows again and again how 'first' technology – tractors, irrigation pumps, chemical fertiliser, pesticides, mills for crop processing – are captured and largely monopolised by the 'haves'. In contrast, 'last' values and technology are closer to the poorer rural people and serve them better. But their neglect is systemic; and an optimal balance between 'first' and 'last' can only be achieved if these biases are reversed through a vast number of personal choices and actions.

'First' values are, however, deeply entrenched. Derived from and appropriate for the rich, powerful, industrialised and heavily armed world, they are carried from the centre outwards in many ways – in textbooks, in hardware, through the media, through consultancy. For reversals towards the rural poor, big obstacles lie in the citadels of professional purity in the metropolitan cores. Sustained by and supplying the needs of the 'first' world, they are subject to little demand or influence from the 'last', third, world, let alone from the rural poor. One example is the policies and priorities of those custodians of professional values, the editors of journals. They are faceless but powerful. They influence what is written and disseminated, and the content and style of research. Moreover, academic appointments boards all over the world, examining the curricula vitae of candidates look at their lists of publications. More weight is given to publication in journals which are 'international', that is, based in the industrialised countries, than in journals which are 'national', that is, based in third world countries. More weight is also given to publications in 'hard' journals which are believed to have rigorous standards of acceptance according to strict professional norms, than in 'soft' journals which may be more wide-ranging, more inter-disciplinary, and more original.

This discourages imaginative and inventive rural research. A third world student in a rich country university wished to do his PhD on ethno-soil science – to study rural people's knowledge of soils, and its relationship to modern scientific knowledge – but he was dissuaded by his thesis advisers who said it would be bad for his career since he would be unable to publish articles in any of the 'hard' journals. A rich country professor working in a third world country was asked to devise new statistical methods for agricultural research on inter-cropping as practised widely (and rationally) by small farmers in that country; but the papers he then wrote were rejected by the international journal which he had formerly edited. Agricultural researchers in a third world country were reluctant to collaborate with those developing new methods of learning from farmers because they feared they would be unable to publish the results. In another third world country, scientists who worked in villages, devising appropriate technologies jointly with villagers, found that their institute's own journal would not publish their work. The supposed or actual policies of journal editors can thus undermine or deflect sensitive rural research. There are signs of shifts; but the assault on journals must be sustained and intensified if a fairer balance for the 'last' is to be achieved.

This requires reversals of values both in rich country cores and in the cores of the third world. In the rich countries, there are encouraging indications. There are universities and university departments, where the 'last' attributes are valued, and deliberate attempts made to offset biases towards 'first' attributes. Third worlders who go to these universities may now be exposed to a more reversed and balanced set of concerns and values than a decade ago, or than some would easily find in their own countries. In agriculture, the International Agricultural Research Centres have helped to raise the status of work on poor farmers' animals and crops – including ILCA's[4] work on goats, IRRI's on pest and disease resistance in rice, and on rain-fed rice, CIAT's and IITA's on cassavas and yams, and ICRISAT's on pearl millet, sorghum, chickpeas and pigeonpeas. In medicine, schools and institutes of tropical medicine and health have been prominent in supporting new 'last-first' approaches. Paradoxically, shifts of values have often proved easier and quicker in the rich than in the poor world; and this means that some rich country professionals are now well placed to help colleagues in third world institutes and universities who are working for change but who are impeded by hierarchy, rigidity, and reaction.

Bastions of conservatism remain in all countries. Reversals threaten established distinguished figures, eminent in their fields, respectable and respected arbiters of orthodoxy. In rapidly evolving fields like nutrition, irrigation management, and social forestry, opposition can be expected to the threat of new ideas. Even more, the idea that farmers should be teachers will be resisted. In one Third World university, an MA student with an imaginative supervisor was set to work to learn from farmers as his instructors. When the time came for the public examination, the farmers were asked to list the questions which should be presented to the candidate. The farmers came to the oral, dressed in their best clothes. After their searching questions had been put, the official examiners saw no need to ask any more, and the candidate was passed. But the incident offended the sensibilities of the university establishment, created a furore, and was never repeated.

One hope lies in the comparative, but unrecognised, advantage of Third World professionals. A researcher in, say, MIT, Reading or Paris is far from the rural Third World; but the researcher in Bamako, Bangalore or Bangkok has it closer. Moreover, Third World scientists working on 'high' technology are repeatedly trumped by Western scientists who have the advantage of less hierarchical research organisations, more resources, and more accessible and rapidly evolving support services. The 'first' values of Western technology draw many able Third World scientists away from the opportunities which surround them. Some, like Amulya Reddy and the ASTRA[5] group at the Indian Institute of Science at Bangalore, have turned around and begun to show what can be done. Further, for their work in villages they have received more recognition than they might have done for conventional 'first' research. But thousands of others are pointed in the opposite direction, towards work where they are at a disadvantage, and where it is difficult or impossible, however brilliant and able they are, to be ahead in their fields and to make useful additions to knowledge unless they emigrate. Ironically, in few spheres are such additions to knowledge so easily, cheaply and simply made as where professionals are most programmed not to seek them: in learning what rural people know that researchers do not know. With 'last' work generally, past neglect means future promise. It is with 'last' work that Third World researchers have a comparative advantage and can most readily trump those of the rich world. For many of them, it is by turning outwards towards this accessible frontier, the poor periphery, that the gains, both for themselves and for rural people, can be greatest.[6]

Reversals into gaps

Specialisation is a parallel problem to the attractions of 'first' values. It both unites and separates the two cultures – of academics and practitioners. The 'first' links between the two cultures are stronger in the physical and biological sciences – geology, hydrology, engineering, soil science, agriculture, veterinary and animal science, medicine, and forestry, for example – than in the social sciences. But with all sciences, it is in the academic culture that specialisation tends to be more marked. The practical world demands breadth, intelligibility and usefulness, while academics are freer to dig deeper down into narrow ruts, so often in the process becoming esoteric and obscure. Each discipline develops its own concepts, jargon and priorities. Disciplines become inbred and spawn subdisciplines, beyond the understanding of lay people. Soil scientists, we are told, are divided between the pedogenetic and edaphological approaches (Moss, 1979). We learn that there has been a failure to connect between ecological anthropologists and lexicographical anthropologists and that unfortunately 'the demands of both fields have left little room for crossing subdisciplinary lines' (Brush, 1980, p. 38). Hyper-specialisation may be intellectually exciting and sometimes useful, but it can also be blinkered. The old jibe says that people know more and more about less and less; the corollary is that they know less and less about more and more.

To counter the ignorance inherent in specialisation, one common remedy is multi-disciplinarity – the adding of disciplines to disciplines. The assumption is that for any purpose – such as research, project appraisal, monitoring or evaluation – all relevant aspects will be well covered if enough of the available disciplines are mustered. There is much justification for multi-disciplinarity, and it can be especially fruitful when social scientists and physical and biological scientists manage to combine well. But there are also difficulties and flaws in it. Sometimes the more the disciplines, the more cautious people become not to trespass on the territory of others, and the more they focus the beams of their searchlights to shine brighter on smaller patches which they can safely claim as their own. The more the disciplines and the larger the team, so too the more difficult it becomes to communicate and to integrate work. In the report, specialised sections are tacked together. The 'hard' technical parts (Chapter 3: 'Soils'; Chapter 4: 'Hydrology') come first and the 'soft' human parts (Chapter 12: 'Sociological constraints'; Chapter 13: 'Impact on Women') come as an unconnected residual, last. Far from illuminating everything, a full battery of disciplinary searchlights may serve only to dazzle and confuse.

In practice, also, gaps are left in both analysis and action. These include three forms: first, there are gaps between the disciplines; second, gaps occur because although the disciplines or professions exist, they are not represented in the rural scene; and third, there are gaps in modes of analysis. All three present opportunities for serving better the interests of the rural poor.

Gaps between disciplines, professions and departments

Many gaps can be discerned between disciplines, professions and departments as they are conventionally oriented and organised. Illustrations can be taken from biology, from energy, and from irrigation management.

In the biological sphere, neglected areas can be found in various linkages between crops, animals, fodder, trees, groundwater and fish. These have often fallen between disciplines and between departments. Agriculturalists are concerned with field crops, not tree crops. Animal specialists are frequently more concerned with veterinary science and animal health than with animal nutrition. Foresters are concerned mainly with protecting trees in forests and with commercial plantations, not growing them on farmers' fields. Hydrologists and irrigation engineers look for physical and mechanical solutions to problems. Pointed and conditioned in these ways, the professions are poorly equipped to recognise and exploit big opportunities.

To be specific, integrated animal and crop husbandry, despite the many linkages between draught power, crop residues, fodder, manure, and crop production, has fallen between agronomy and animal husbandry (McDowell and Hildebrand, 1980). The use of tree fodders and tree fodder intercropping, which now promises big advances in rural productivity, has fallen between forestry, agronomy and animal husbandry. The use of trees as biological pumps to tackle waterlogging has fallen between forestry, agriculture, hydrology and irrigation engineering. Fish ponds have fallen between fishery departments concerned more with large-scale and marine fisheries, and agriculture and forestry which might provide feed for the fish. Crop wastes (the word 'wastes' begging the question) have been neglected by all. In these examples, the built-in specialisation, conservatism and rigidities of university teaching, research institutes, and government departments point away from the opportunities.

In the energy sphere, new energy sources and technologies also present gaps. Ecological energetics – the study of biological energy flows and efficiencies (Phillipson, 1966) – has been a slow and late comer to rural development, not least because quantified research is laborious and has to be painstaking. Yet the energy crisis since 1973 has opened up huge potential for poor rural people. Some of the crops of the poor, such as cassava (manioc, tapioca, yucca) are energy sources, and can be used to make alcohols. Wastelands which previously appeared of little use except for communal grazing now acquire a new value as potential energy plantations; and a whole movement of social and community forestry is gaining momentum in India, the Philippines and elsewhere, with the possibility that the poorer people may be the main beneficiaries if the passing chances are seized. Producer gas technology, by which cars, irrigation pumps, and other machines can be powered, is coming back into its own, neglected since the Second World War, and may generate markets for fuel gathered or grown by the poor and sold by them.

It is a sad sign of inertia or lack of imagination that the Consultative Group for International Agricultural Research (the body responsible for IRRI, ICRISAT, CIMMYT, etc.) has not set up a research institute with as its central concern the development of biological and other new energy technologies to benefit poorer rural people. A great chance has been missed. Nor have governments seized this opportunity. The best work has been done by voluntary organisations, inspired individuals, and small breakaway groups.

In the sphere of irrigation management, canal irrigation in South and South East Asia and elsewhere presents a similar gap between disciplines, professions and departments. The management of main irrigation systems, including the scheduling of water deliveries, falls between irrigation engineering, agronomy, agricultural engineering, and sociology, and is the prime concern of none. The civil engineers in charge of canal irrigation systems are professionally interested in design and construction (Jayaraman and Jayaraman, 1981) for which they have been trained, less interested in maintenance for which they have less training, and least of all in the operation of canals and the distribution of water, for which they are scarcely trained at all. Their attention is mainly at the headworks and the major controls. For their part, agricultural engineers are engaged only at the lower levels, below the outlets, close to and on farmers' fields. Agronomists too are preoccupied with the field level where crops are grown. Sociologists also concentrate low down, at the community level. There is no profession for which the management of the main canal system is a priority. It is a blind spot (Wade and Chambers, 1980). Farmers privileged to be at the headreaches receive abundant water while those deprived at the tail receive little or none, and receive it unreliably and late. The opportunities for improved equity and productivity are immense (Bottrall, 1981), but are not seized, not least because of the design and construction biases of the training of civil engineers, and the common chasm between Departments of Irrigation and of Agriculture.

Missing disciplines, professions and departments

Other gaps are left because disciplines, professions and departments are not represented, or are very poorly represented, in rural development. Two examples are management and law.

Management is a discipline or profession which has yet to make its major contribution to rural development. Overwhelmingly, it has had an urban, industrial, and commercial character. Academics have come towards rural management through studies in public administration, though with a persisting reluctance to see procedures as having intellectual as well as practical interest. Practitioners have come towards it through hard experience and improvisation and inventiveness to meet problems and needs as they have arisen. There are now quite numerous studies which combine empirical analysis with practical prescription.[7] Institutes of Management which include rural and agricultural management have been set up, and in India there is an Institute of Rural Management in Anand. All the same, a body of practical theory for rural management is only now beginning to emerge, and it is rare indeed to find people who would describe themselves as specialists in rural management. Rather few academics see rural management as intellectually exciting or as a realm for action research; and practitioners still have a long way to go in recognising how management can contribute to rural development.

Law is a profession which, like management, has a strong urban, industrial and commercial orientation. One does not find a rural Legal Department equivalent to a Department of Community Development or Agricultural Extension. Yet there are many laws in many countries which, if enforced, would help the rural poor. As it is, the laws of property, invoked by the 'haves' against the 'have-nots', maintain and defend gross disparities of wealth. In famines, as Amartya Sen has argued, the law stands between starving people and food: he concludes his book *Poverty and Famines* (1981) with the words 'Starvation deaths can reflect legality with a vengeance.'

The need is to reverse this tendency: with patience and courage, step by step, to invoke the law to enforce measures which should favour the poor – to enforce land reforms, the payment of legal minimum rural wages, the access of the poor to services and programmes designed for them, and low interest rates from moneylenders, and to prevent illiterates being cheated into signing documents accepting ruinous interest rates for debts, surrendering their land, or forcing them into bonded labour. Legal aid for the rural poor has a history of isolated successes (e.g. Bagadion, Espiritu *et al.*, 1979; Mehta, 1979). But in most places it is a gap, a void crying out to be filled.

Neglected modes of analysis

Modes of practical analysis were discussed in Chapter 6 – in terms of costs and choices, of causes and constraints, of finding and making opportunities, and of political feasibility. To these may be added two more which touch matters critical for poor rural people.

The first is seasonal analysis. The tropical wet seasons are a gap. Urban-based professionals visit rural areas most during the dry seasons when travel is easiest and most congenial and people are at their healthiest, happiest and best fed, and least during the rains when travel is difficult and uncomfortable, people are most liable to be sick, under stress, and short of food, and ratchets of impoverishment are likely to be most common. Wet season interventions are therefore neglected, and wet season conditions under-represented in rural analysis and planning. Yet there are many opportunities in this gap: organising seasonal crèches for children while their mothers work in the fields; stocking clinics with medicaments according to seasonal needs, and ahead of breaks in communications with the rains; developing cropping systems to produce food earlier after the start of the rains; seasonal credit; and so on.[8] Such measures are found, but are still far too rare.

The second mode is to reverse analysis, from top-down to bottom-up. The top-down mode starts with disciplinary specialisation and uses its tools to examine rural situations. Bottom-up analysis starts with the condition of poor people, their resources, aspirations and problems. It might better be described as within-poverty analysis, for it entails trying to see from within, to adopt a diametrically opposite world view in order to see what might and ought to be done. At once, this will give a more holistic view, cross-cutting outsiders' disciplines, than the top-down approach. It is likely to identify needs and gaps which, interacting with specialised knowledge, show new opportunities. Moreover, just as with seasonal analysis, so with analysis which starts from the concerns, world views and knowledge systems of rural people, outsiders are provided with new frameworks of categories which may make their inter-disciplinary collaboration more feasible, more exciting and more useful.

Conclusion: gaps as centres

Often for the poor, gaps are central. The resources and developments which are main line concerns of established disciplines, professionals and departments are usually linked to and taken up by commercial interests and by those who are better off. Priorities are set by conventional specialised analysis and influenced by political forces which usually favour the strong. Opportunities for the poor lie precisely where resources have been protected by 'first' biases and by the narrowness of specialisation. It is here that new economic niches and new livelihoods can be generated by exploiting slack resources and using new technology. A great practical challenge is presented by the proviso that during critical periods of competence (pp. 158–9), poor people must be enabled to establish secure rights to the resources and the flows of income from them.

The new professionalism that is sought will, then, reverse tendencies to exclusive and increasing specialisation. There will always be a case for highly trained professional competence and for rigorous research. Nothing here should be construed as an attack on that. But many of the better chances for the poor lie elsewhere, and can be found and seized through wider and more open-minded observation, discussion, learning and analysis than that of any one discipline, profession or department. Narrowness among outsiders is a luxury poor people should not be asked to afford. Professionals should neither confine themselves to their own disciplinary territory nor fear to trespass in that of others. If they are to see the gaps and help the rural poor to exploit them, outsider professionals have to be explorers and multi-disciplinarians.

Political economy for all

Political economy is more a set of questions than a discipline. In practical rural development it is concerned especially with who gains and who loses. These questions have usually been left to social scientists. But they are too vital and too universal to be confined. One high cost of the gulf between the two cultures is that these questions are asked mainly by negative social scientists and much less by positive physical and biological scientists and practitioners. This needs correcting; political economy is too important to be left to the social scientists.

There are two reasons for this: first, who benefits is affected by many decisions which appear technical or neutral; second, awareness of who gains and who loses is a precondition for realistic interventions to benefit the poor.

On the first point, the technical and neutral appearance of many decisions and actions is deceptive. A decision by a bureaucrat or politician to instal a Modern Rice Mill in a South Asian country can be presented as entirely technical, based on considerations of milling out-turn efficiencies, benefit-cost ratios, and the like. But Barbara Harriss has calculated that running at effective capacity, one such mill in South India would put some 300 people out of work (1977, p. 295), many of them very poor and vulnerable women. In such circumstances, an aid agreement for a Modern Rice Mill can be a death warrant. Those who sign such warrants 'know not what they do'. They make 'technical' decisions, lacking the knowledge, insight and imagination to realise what they will mean to the distant, weak, silent, poor. But the reality is there. The causal chains follow through from the ink on the agreement to the poor woman who once made do but who now finds no work, no money, no food, and no hope. So much negative social science has failed here; concerned more with class, quantification, and the macro-level, it has rarely laid bare the personal detail of these causal chains in their blind awfulness.

Or again, as is now much better recognised than a decade ago, decisions taken in agricultural research affect who benefits. Larger and more prosperous farmers can afford and obtain fertilisers, pesticides, irrigation water, and hybrid seeds. To many smaller and poorer farmers these are out of reach. Agricultural scientists have often regarded production as an end in itself, but they increasingly recognise that *who* produces is at least as important as *how much* is produced. Many decisions in agricultural research affect the *who*: decisions to work on biological nitrogen fixation (available to more, smaller, farmers) as against responses to chemical nitrogen (available to fewer, larger farmers); or work on inbred biological resistance to pests as against pesticides, or on crops and varieties sensitive or resistant to stress from water shortage, which is more likely to occur in the fields of the poor. Such decisions affect who produces how much and with what risks, and thus the food supplies and incomes of millions of poor people. If the scientists who make such apparently technical and neutral decisions could always envisage the ramifying effects, over the years, of their choices, they might be astonished and appalled at the power they have to give or to take away.

Examples could be multiplied to include decisions about R and D in other fields, in other choices of technology, and in elements in the design and implementation of programmes and projects. Moreover, there are many other decisions – on devaluation, on pricing policy, on regional development, on the siting of infrastructure – which more obviously affect the rural poor and who gets what. The decisions as between tertiary, secondary and primary education; between resources for central hospitals or for peripheral primary health care; between capital-intensive or labour-using techniques for building and maintaining roads – are now frequently debated; a mark of progress over the past two decades.

But many decisions critical for the rural poor remain hidden and lost in the dark and labyrinthine places of government. The general point can be illustrated with one instance. It is taken from Zambia, but examples could be found in any country, rich or poor, in the world. In Zambia in 1980 it was reported that women in the remote Chambeshi Valley were doing a day's work threshing paddy by hand for a payment of about 100 grams of salt. At the nationally controlled price for salt, this was equivalent in value to about one-hundredth of the statutory minimum urban wage (ILO, 1981, p. 23). But there was an endemic salt shortage in rural areas, so that its black market value to the rural poor was very high. Now Zambia's salt is imported and requires a foreign exchange allocation. If an inadequate foreign exchange allocation was the reason for the salt shortage, the question is how this could occur. There are several possible explanations: the political impotence of peripheral people; corruption, since the black market must have been very profitable to some; muddle and inefficiency – getting the arithmetic wrong, or implementing decisions slowly; allocating cuts where the political price would be lowest; sheer lack of awareness; or some combination of these. Had the officials and others concerned been exercising their imaginations, and asking who gains and who loses, they might have seen that it would be the poor who would lose. There is also the awful possibility that they were aware, saw this, and still acted as they did, preferring Mercedes Benzes for the few to salt for the many.

The second reason why the questions 'Who gains? Who loses?' are too important to be left to social scientists is more tactical. The discussion of political feasibility (pp. 160–3) used the image of the brick wall of political will, and explored different distributions of benefits between groups. This suggested two types of approach: those which are piecemeal and oblique and those which are frontal.

Most rural development plans, programmes and projects intended to benefit the poorer people are non-frontal. Both 'spread-and-take-up' and 'last – first' interventions can quite often enable the poor to benefit without constituting a direct threat or direct loss to the rich. Depending on local conditions, a school, a health post, a road, may be to the advantage of all. This is the Chinese 'all boats float higher', a game where all gain, even if some gain more than others.

The use of natural resources can also be analysed to search for ways in which all can end up better off. The tendency is to regard land, water, forests, fish and grazing as resources to be appropriated so that one household's gain is another household's loss. This is by no means always so. With canal irrigation water, scheduling is sometimes possible so that all gain: those at the top end gain by using less water delivered in a more reliable, timely and controllable manner (with less flooding, less waterlogging and less consequent salinity), while others at the tail end gain from having more water also delivered in a better manner (Valera and Wickham, 1979; Early, 1980; Chambers, 1980; Bottrall, 1981). With common property natural resources – common land, forests, fisheries, and grazing lands – similar questions arise. On the one hand there are dangers of appropriation by the powerful: large-scale enclosers with common land, timber contractors with forests, trawlers with fisheries, and 'big men' (owners of large herds or flocks) with grazing. On the other hand, there is the great challenge to overcome the tragedy of the commons and so manage these common resources that production will be maintained and increased in ways which enable all to gain.

The search for non-frontal approaches like these should not undermine those who work for frontal approaches where they are needed. Nor should it weaken those who seek to push benefits more and more to the poor where the less poor must lose. It is an argument, rather, for all professionals concerned with rural development to worry about and think through the implications of their decisions and actions, and for all – politicians administrators, magistrates, lawyers, agricultural scientists, foresters, community developers, health staff, veterinarians, animal husbandry specialists, educators, and others – to shift their sights and priorities from rich to poor; to ask, again and again and again, what the effects of their decisions and actions will be on those who are poor, physically weak, isolated, vulnerable and without power.

New professionals

These arguments make a case for a new professionalism of putting the last first and for new professionals to develop and practise it. This could be dismissed as an Alice in Wonderland fantasy, or as an unattainable saintly ideal. But hard experience shows that it is possible, that there is space in which to move. New professionals already exist. They are those whose choices of where to work and where to allocate resources and authority reflect reversals towards the periphery and the poor; whose analysis and action pass the boundaries of disciplines to find new opportunities for the poor; and who test policy and action by asking who gains and who loses, seeking to help those who are deprived to help themselves. They are those who recognise small farmers, artisans, and labourers as fellow professionals and set out to learn from them. They are those who abandon disciplinary boundaries, and those who span the two cultures of academia and practice, taking the best from each – criticism from the one, and vision and action from the other. They are those whose values and actions put the last first.

It is easy to write about what ought to be. The hard question is how, in the real, messy, corrupting world to encourage and enable more people to move in these directions; how to multiply the numbers of committed outsiders – politicians, government staff in the field and in headquarters, voluntary workers, religious leaders, researchers, teachers, trainers – who see the need to put the last first, and how to stiffen their courage and will to act.

Notes

1 To the best of my knowledge. There is a *Journal of Peasant Studies* but I do not think that is quite the same thing. The *Economic and Political Weekly* in India publishes many articles on rural poverty but this is only one of the subjects it covers.

2 See Chapter 4, pp. 81–2. Goat droppings are, however, predictable within limits, and visible, even if they often pass unseen.

3 See for example Byres, 1972; UNRISD, 1974; Palmer, 1976; Dasgupta, 1977; Farmer, 1977; Hameed *et al.*, 1977; Pearse, 1980; Harriss, 1982.

4 Respectively the International Livestock Centre for Africa in Ethiopia; the International Rice Research Institute in the Philippines; the Centro Internacional de Agricultura Tropica in Colombia; the International Institute of Tropical Agriculture in Nigeria; and the International Crops Research Institute for the Semi-arid Tropics in India.

5 The Cell for Application of Science and Technology to Rural Areas.

6 For an outstanding illustration of this point, see Leela Gulati's book *Profiles of Female Poverty*.

7 For example, Hunter, 1970; Uphoff and Esman, 1974; Lele, 1975; Leonard, 1977; Hunter, 1978; Korten and Alfonso, 1980; Korten, 1980; Esman and Montgomery, 1980; Moris, 1981.

8 For longer lists of counter-seasonal interventions see Chambers, 1980, and Chambers, Longhurst and Pacey, 1981.

CHAPTER EIGHT

Practical action

That low man seeks a little thing to do,
    Sees it and does it:
This high man, with a great thing to pursue,
    Dies ere he knows it.
That low man goes on adding one to one,
    His hundred's soon hit:
This high man, aiming at a million,
    Misses an unit.
        Robert Browning, *A Grammarian's Funeral*

Although the freedom and power of individuals vary enormously, all can do something. One starting point is to question values that put the last last, such as the belief that the poorer and weaker people are, the less they should be paid. Another is to exercise imagination in thinking through the distant but real effects on the rural poor of technical and policy decisions and of outsiders' actions and non-actions. Yet another is to increase contact with and learning from the rural poor, offsetting the anti-poverty biases of rural development tourism, and using methods of rapid rural appraisal. Reversals in learning can take many forms, including sitting, asking and listening; learning from the poorest; learning indigenous technical knowledge; joint R and D with rural people; learning by working; and simulation games. Reversals in management entail changes from authoritarian to participatory communication; fewer staff transfers and better staff in poor and remote districts; and enabling and empowering weak clients to make effective demands for services and for their rights. To achieve reversals, it is best to start by acting and learn by doing. Most of the time, the best way forward is through small steps and little pushes, putting the last first not once but again and again and again. Small reversals then support each other and together build up into a greater movement. Many already put the last first, some from a distance, and some directly. The hope of this book is to support all of them, and to encourage others to do what they can.

The scope for personal choice

Some social science tends to minimise the scope for personal choice. Negative social science (pp. 30–33) does this in two ways. First, it stresses how awful things are, and how they are worse than they appear. This emphasis has often been salutary. By repeatedly pointing out how rural elites act as a net to intercept benefits, how poorer people are exploited, how capital-intensive technology can destroy the livelihoods of the poor, how women so often suffer, how transnationals exploit poorer countries, and so on, social scientists have exposed scandals and myths and helped to free practitioners from delusions. But negative social science can be accompanied by a sort of fatalism in which the worse things are the less hope there seems to be for doing anything about them.

Second, social science seeks uniformities and laws. While practitioners get on with the detail of doing things, some social scientists try to generalise through theory – the laws of capital and of the market, and the analysis of class, dependency, comparative advantage, supply and demand, and the like. Such theory and generalisation can lead towards determinism, postulating forces which can imply individual impotence in the face of great historical movements. The awfulness can then appear inevitable and those who seek to achieve change can seem deluded.

Some negative social scientists may now be waiting for the next sentences which, in a now well-established tradition, would call (lamely) for the (improbable) exercise of political will to effect the desired reversals; and would see this call as yet another example of a naive reformism which calls upon and expects the wealthy and powerful to act against their interests.

The argument here is different. It is based on the premise that individual behaviour is not fully determined, a premise on which all, including determinists, appear to base their lives. Political, social and economic forces do operate; but when they are dissected, sooner or later we come to individual people who are acting, feeling and perceiving. They may be women or men, poor or rich, rural or urban, illiterate or educated, from less developed or more developed countries; landless labourers, artisans, small farmers, workers in voluntary agencies, government field staff, prosperous farmers, traders, workers, business people, politicians, senior officials, aid agency staff, professionals, diplomats, international bankers. All are found deciding what to do and all are to some degree capable of changing what they do.

What varies is the scope they have to act differently and the extent to which those different actions can change what happens.

This depends on their freedom and power. Poor rural people have little of either: a landless labourer locked into a single 'hedgehog' relationship or a 'fox' busily struggling to survive, has little room for manoeuvre or influence. The President of the World Bank, by comparison, has vast scope. And in between these extremes there are outsiders with varying degrees of freedom and power. For most of them, most of the time, the opportunities are for small steps and little pushes. To the individual these may seem insignificant, but the sum of small actions makes great movements.

Many such small actions are indicated by the themes and detail of earlier chapters – needs to perceive rural poverty and to understand its nature; to span the three cultures of academics, practitioners, and rural people; to devise and use cost-effective ways to find out about rural situations; to sit down, listen and learn from rural people; to load the dice differently for the rural poor; to think well in seeing what to do; and to foster a new professionalism which asks who gains and who loses and which puts the last first. Each of these presents an agenda of action. The question is who should and can start where and how, and to do what.

For outsiders, the answer is that all of them can do something.

Those in the rich world are spatially remote from the rural periphery but are part of the global system which maintains rural poverty. They can help themselves and others better to understand the issues and linkages; they can lobby for better terms for Third World trade and for better programmes of aid; they can support campaigns to stop abuses which harm and impoverish those who are already poor, like the export to Third World countries of drugs which are dangerous or needlessly dear; they can support voluntary agencies which take direct and effective action. Academic researchers, through their choices of topics and of methods to investigate them, can illuminate processes of enrichment and impoverishment, the patterns of power, ignorance and prejudice, the nature of rural deprivation, and successes and failures in rural development. Those who work in aid agencies can argue and work for programmes and projects which help the poor and against programmes and projects which harm the poor. They can ally themselves with and support those in host country governments who share their aims. Rich country professionals, whether in the medical, engineering, physical, biological or social sciences, can question their values and the influence which they have outside their immediate rich country environments.

And always, at its simplest and most direct, with OXFAM, Christian Aid, Catholic Relief Services, War on Want, Save the Children Fund, Terre des Hommes, and other international voluntary agencies which work on rural development with and for the poor, impact on rural misery is no further away than the cheque book or money order and the post office.

For Third Worlders, closer to the action, the opportunities are more obvious but the difficulties greater. This final chapter is for all outsiders, including those from rich countries, but most of the points try to address the predicament of that great majority of rural development professionals who are nationals in Third World countries. These include especially the field and headquarters staff of government organisations, parastatals, voluntary agencies, and organisations in the private sector, and the staff of universities and research institutes. Many of them feel trapped. Others feel that their efforts against enormous obstacles and powerful forces will be of no avail.

What can they do?

It is easier to write about what to do than to do it. Writing does not require courage, but courage can be needed for action. Though much rural development, especially in Africa, is welcomed by the whole population and does not involve outsiders in personal risk, much also, especially in Asia and Latin America, involves conflicts of interest where the weak are dominated, exploited and cheated by the powerful. Where that happens, many of the rural poor and those who work with and for them face abuse, discrimination and danger; the bravest and most direct are often threatened; some are assaulted; and some are killed. However much such people should be applauded and honoured, I doubt whether foreign outsiders like myself, protected by passport, police and the state, can be justified in urging others to risk their livelihoods, their families' well-being, or their lives. To take risks for oneself is one thing. To encourage others to do so is quite another. Even more, for any outsider to encourage vulnerable poor people to take risks raises ethical questions, especially when it is they, and not the outsider, who will pay the price of failure. The most honest and painstaking reflection is demanded first.

There are anyway many degrees of negotiating force and pressure short of physical violence, and skill and self-control can be exercised to choose the best level. It has to be recognised, though, that where exploitative interests are entrenched, force and pressure which are succeeding against them can trigger violence from the other side. Nothing that follows should undermine those whose justified non-violent acts against

193

exploitation provoke abuse and violence against themselves. From those of us who are far from the sharp end of rural development, they deserve respect and admiration.

At a less dramatic level, action can take many forms. My hope in this chapter is that by showing several sorts of straightforward action, more people will be encouraged to do more; to add their steps and pushes, however small, towards those great reversals which are needed; and that those who read this will find at least something that will make sense to them, that they can do, and that will work.

Beliefs, values and imagination

Action and introspection go together. It is often best to start with action. But to put the last first confronts basic beliefs, challenges personal values, and demands the exercise of imagination. Beliefs and values here refer to what an outsider considers right and fair for the rural poor; and imagination refers to the willingness and ability to think out and recognise the effects of outsiders' actions and non-actions on them. To support action, a reversal of beliefs and values is often required, a sort of 'flip', a switch to seeing things the other way round, from the other end.

Take, for example, the basic belief that the poorer people are, the less they should be paid. It is important here to distinguish the factual statement, that the poorer people are the less they *are* paid, from this moral statement, that the poorer people are the less they *should* be paid. The factual statement is true; but when the objective is to reduce poverty, the moral statement is starkly perverse, since the more poor people are paid, the less poor they become. To say that poor people should be paid less is to say that they should stay poor. Yet whatever private and public views are expressed about reducing and eliminating poverty, the perversity persists. It is almost universal. Women are paid less than men; the disabled receive less than the able-bodied; the illiterate get less than the educated.

Nor is this just a straight economic question of paying for results, or for value added. There are deeper, ingrained attitudes. Here is an official's opinion about payments to people in India, mainly Tribals, who pick *tendu* leaves for making *bidis* (local cigarettes):

> The present wage rate (1 paisa for 20 leaves) is not remunerative. But then it is not meant for able-bodied persons. It is to be appreciated that tendu provides seasonal

employment to women and children or to the old. It provides employment to those who otherwise will not be able to get employment elsewhere. Thus, whatever he or she earns, should be accepted as fair.

(quoted in Sarin and Khanna, 1981)

Should it be accepted as fair that poor people who have no alternative employment should be paid an unremunerative wage? Can the answer be anything but no? Yet practices which follow this principle are very widespread. There are, it is true, public works programmes which pay minimum wages, and among which the Employment Guarantee Scheme in Maharashtra in India (Reynolds and Sundar, 1977) stands out as an example. But in spite of such exceptions, the perverse principle of 'pay the poor less' is deeply rooted, both as an axiom of the market and as a precept of the privileged.

That the principle is applied in the private sector is only too well known: some employers pay pitiable wages when labour is in surplus and when they see this serving their private interests. Less well recognised, the same principle prevails in governments, including governments which profess to be at war with rural poverty. Reinforced by urban biases, the principle becomes 'pay the *rural* poor less'.

The point can be illustrated by government-regulated producer prices in Zambia. As Table 8.1 shows, between 1971 and 1979, the prices for four rural products all rose less than the low-income urban price index, suggest a shift in the terms of trade against rural producers; and among these four, maize (the major product of the larger and commercial farmers) almost kept pace,

Table 8.1 Changes in some rural producer prices in Zambia, 1971—79

	1971	1979	Percentage change 1971—9
Low-income urban price index	100	242	142
Maize	4.00	9.00	125
Dried fish	31.00	50.00	61
Charcoal	1.75	2.50	43
Sorghum	4.70	6.00	28

Source: ILO, 1981, p. 65.

while the marketed products of the poorer rural people – dried fish, charcoal, and sorghum – lagged far behind, indicating how government action or neglect made poorer those who were already poor.

The producer price for honey in Zambia is an even more striking case. In 1980, some 70 per cent of the population of the remote and poor Kabompo District were believed to be beekeepers and the dispersed population had little to sell but its honey. But the Forest Department paid them only 0.41 kwacha per kg, a mere 12 per cent of its selling price of 3.50 kwacha on the Copperbelt; nor could this be justified in terms of intermediate costs, since almost all processing, transport and other expenses were paid for through a German aid programme. When pressed to raise the price paid to honey-gatherers, an official replied that this could not be done unless there was evidence that production costs had risen, as though such a question was relevant where government was already exploiting poor people so much, and so much more than any private trader might have dared. It is in ways such as these that the urban and class reflexes of bureaucrats keep poor rural people poor. They can use such arguments with impunity. The rural poor are, after all, powerless. The rich would never stand for it: but then they do not have to.

These examples are both distressing and hopeful: distressing because of the mindless insensitivity which they expose; hopeful because ignorance, lack of imagination, lack of feeling, and urban and class reflexes on the part of those who make decisions present an opportunity for change. It is true that on one level these are questions of organisation and power: in Zambia and elsewhere, urban workers and commercial farmers are well organised as pressure groups, while no one may represent the interests of scattered, peripheral and uneducated fisher-people, charcoal-burners, sorghum-growers, honey-gatherers and women labourers. But it is also true that the interests of these poor groups can be represented in many ways by those who are powerful. To reverse the trends against the rural poor requires that more and more of those powerful people know, care and act. In the case of the producer price for honey, perhaps all it needed was for some key person, or some coalition, to realise what was happening and to argue on paper, in committees, and in the press that the publicly proclaimed objectives of helping the rural poor could be admirably served by raising the price. When other government programmes were struggling to improve the lot of the poor, here was an opportunity to augment the incomes of precisely those who were hardest to reach, with a minimum of administration and a maximum of self-reliance on their part. The opportunity could have been seized if a few people had exercised their imagination, cared and acted.

Imagination is vital. Outsiders need to envisage the distant but real and ramifying effects of their actions and non-actions, thinking through the causal chains which flow from them. The research of social scientists could help more here: it has rarely traced through such centre-outwards linkages in the human detail of case studies, what Stephen Biggs has called 'slice' research. But even without the empirical description of the personal detail of these links, reflection can suggest the connections: between the decision to install a Modern Rice Mill, displacing poor female labour, and the resulting misery of widows and women heads of the household and mortality among their children; between the low priority which courts give to cases brought on behalf of the poor, and the inability of the poor to stand up against exploitation; between the decision to breed crop varieties for ideal controlled conditions and the lack of benefit to resource-poor farmers; between the rapid transfer of officials in peripheral areas and the ineffectiveness of anti-poverty programmes there; between the allocation of low import quotas to soap and the scourge of scabies; between building big hospitals while failing to mount a programme to popularise oral rehydration, and the avoidable suffering and deaths of children; between denying vehicles and fuel to staff in remote areas with scattered and deprived populations, and the resulting paralysis of services; between the removal of a dedicated official and the rural misery and hopelessness that might have been averted.

The list can be lengthened. Examples are legion. But the point is made: the connections are there; they are real and they are unseen. It takes imagination to recognise them.

What counts, though, is action. Values – putting the last first – and imagination to visualise the distant effects of actions and non-actions, are points of entry. But concentrating on these can be a postponement. It is often best to get on with doing whatever can be done, however small. To help those who want to start or to go further, what follows are some points of entry, for themselves and for others, all leaning in the same, reverse, direction. These actions are designed to be realistic and feasible, bearing in mind human weakness and constraints on personal action. Readers will see more scope for themselves in some than others; but hopefully all will find something which they can use, or which will lead them to yet other actions which put more of the last first.

197

Practical appraisal for outsiders

A first and obvious point of attack is for outsiders to change the ways they learn about rural conditions. One problem here is the tension between the two cultures over time and timeliness: academics inclined to favour longer, unhurried, appraisals; and practitioners needing instant information to meet the deadlines of seasons, budgets, committee meetings, and ministers. As we have seen in Chapter 1, shortage of time contributes to the anti-poverty biases of rural development tourism and to careless and misleading investigations. These problems can, realistically, be tackled by those involved in two ways: by improving rural development tourism; and by developing and using techniques for rapid rural appraisal.

i) Tactics for tourists

Offsetting the anti-poverty biases

For that majority of outsiders concerned with rural poverty who practice rural development tourism, measures can be taken to offset the anti-poverty tendencies of contact. Urban, tarmac and roadside biases can be countered by going further afield and by walking away from roads; project bias by visiting not only projects but other areas and by non-scheduled stops; biases of personal contact by deliberately seeking out the poorer people, by making a point of meeting women, by taking time to see those who are sick at home and not at the clinic, by asking about those who have left or who have died; dry season bias by visiting during the rains, or at least asking about the worst times of the year; the biases of politeness by breaking away from the courtesies and making it clear what is sought; professional biases by trying through introspection to see the limitations of professional conditioning, by learning from other disciplines, by being observant, and by asking open-ended questions.

Spending longer and going further

In many ways the poorer people are at the end of the line. They take the longest to reach; they are the last to speak; they are the least organised, the least articulate and the most fearful. They often keep a low profile. Some are migrants. Women sometimes hide from male visitors. In visits that are rushed, the poorer are those least likely to be encountered. It is after the courtesies, after the planned programme, after the tourist has ceased to be a novelty, that contact becomes easier.

Remoteness is a special problem. Ingrid Palmer (1981, p. 38) has observed that 'Rocketing around in a Land Rover over rough roads for hours on end can be tiring and dispiriting, but there is no other way to reach the furthest and usually poorest villages', and more time must be set aside for reaching them. But the delays and disasters to which rural development tourism is prone prevent or curtail precisely such visits to remote villages and encounters with the poorer people in them. The serious outsider will do well to allow plenty of time in one place, to tour slowly, to spend the night, and to talk and listen after dark. Vehicle and fuel shortages for field staff can be turned to advantage, with a return to the overnight rural stops which were once common in many countries in the days of the trek or safari on foot, by horse, or by slow vehicle on bad roads. Field staff can make more of limited resources for travel by moving less fast, spending more time out, and learning more.

Being unimportant

The cavalcade of cars, the clouds of dust, the reception committees and the protracted speeches of the VIP's visit generate well-known problems. By contrast, the visitor who comes simply, by bicycle or on foot, fits more easily and disturbs and distracts less. Unscheduled visits, walking and asking about things that are seen, planning not to have a special programme, and avoiding the impression of having influence over the benefits which a community might receive, all reduce the dangers of misleading responses and impressions.

ii) Rapid rural appraisal

As we saw in Chapter 3, questionnaire surveys and statistical analysis limit investigations to what can be asked in interviews and what can be counted. The realities of rural deprivation are often missed. The challenge is to question the conventions of academic purity and find better approaches. Rapid Rural Appraisal (RRA) is one family of techniques of cost-effective ways in for outsiders.

These techniques have been widely practised but until recently little written about.[1] They recognise the trade-offs between the cost of information-gathering, and its quantity, accuracy, relevance, timeliness and actual use. Using 'dirty' as a term meaning not cost-effective, they try to avoid both the 'quick-and-dirty' of incompetent rural development tourism, and the pathological 'long-and-dirty' of some questionnaire surveys.

Where time is short they look for an intermediate and appropriate technology, an approach which is 'fairly-quick-and-fairly-clean'.

Many techniques have been suggested. Some are obvious, if neglected. They include searching for and using existing information; identifying and learning from key informants – social anthropologists, social workers, group leaders, university students doing field research, and so on; direct observation and asking questions about what is seen; guided interviews; and group interviews with informal or selected groups. Other techniques are less obvious, like inspecting an area from the air and using aerial photographs. Perhaps most important are the composite approaches, designed for particular purposes such as identifying priorities for agricultural research, like those of Collinson (pp. 67–9), Hildebrand (p. 68), and Rhoades (1982). Other promising approaches are the compilation of family profiles, combining the methods of social anthropology, journalism, and statistical methods (Roberts, 1980); and the approach of BRAC (pp. 72–3) and Swift (1981b), working with rural people themselves as investigators.

Regarding poverty, systematic observation goes a long way. To assess deprivation and identify those who are worst off can also be helped by key indicators. These may show the relative deprivation of an area or group, or pick out those individuals or households who are worst off. Quality of housing is frequently referred to (Honadle, 1979; Longhurst, 1981b; Moore, 1981). Other indicators include wealth in land and animals, and tangible assets such as tools, beds, cooking utensils, and clothing. Anthropometric indicators of nutritional status also have their uses. Or communities may themselves be asked who are the most deprived among them. All these methods have weaknesses, but separately or combined with others they can be used to shortcut lengthier approaches.

One danger with RRA is that it will always be rushed. A corollary of Parkinson's law is that whatever is planned exceeds the time available for doing it. If this occurs, it will once again be the remoter areas and the poorer people who are left out. The opportunity presented by RRA is, by avoiding lengthy methods, so to save and budget time that the poor are let in, as individuals and as families, to be learnt from and understood in more depth. If the tyranny of quantification can be held in check, there is more time to find out about relationships and processes. Techniques of RRA, carefully developed and used, can raise awareness and understanding of rural poverty, and improve actions to attack it.

Nothing here should detract from the importance of sustained, detailed and careful research, statistically rigorous where necessary, and whether in one or several disciplines (pp. 58–9). The argument is rather to devise and use methods which fit resources, problems and needs. As a technology intermediate between rural development tourism and full-scale research, RRA has many applications for outsiders who are serious about rural poverty.

Reversals in learning

Putting the last first means reversals in learning. The litanies of rural developers include 'We must educate the farmers' and 'We must uplift the rural poor'. These can be stood on their heads. Outsiders have first to learn *from* farmers and *from* the rural poor. But many outsiders are hindered from such learning in reverse by their educational attainment, urban status, and roles as bearers and dispensers of modern knowledge. Staff working in rural areas distance themselves from rural people, showing their separate style and standing through clothing, shoes, vehicle, office, briefcase, documents, and manner and speech. Hierarchy, authority and superiority prevent learning 'from below'. Knowledge of one sort perpetuates ignorance of another. Learning has to start at the other end. The farmer must educate outsiders; the poor must bring outsiders down to earth.

Conventional learning through formal schooling, university courses, and staff training can contribute to these reversals through changes in syllabus. One example is to illuminate the problems and rationality of small farmers through the insights which have come out of farming systems research.[2] But for learning in reverse, this is not enough. There are two further methods which deserve to be developed and included in courses and training.

The first of these is learning directly from rural people, trying to understand their knowledge systems and eliciting their technical knowledge. This is still rare as a part of education and training. The second is trying to experience the world as a poor and weak person. The problem here is to enable professionals to step over and see and feel the world from the other end. The humanistic psychotherapies[3] may have methods to offer for this, but their application to the training of rural development professionals has so far been slight.

Reversals in learning can cover many aspects of life and can take many forms. Many more of these should be devised and developed. For the present, any repertoire might include the six which follow.

i) Sitting, asking and listening

Sitting, asking, and listening are as much an attitude as a method. Sitting implies lack of hurry, patience, and humility; asking implies that the outsider is the student; and listening implies respect and learning. Many of the best insights come this way. Relaxed discussions reveal the questions outsiders do not know to ask, and open up the unexpected.

Different approaches are possible, The pooling of knowledge and mutual stimulation of a small casual group can be an excellent source of insight. The composition of a group can also be designed for a purpose. Gill Gordon (1979) reports an effective way of finding out about changes in child-rearing practices in Ghana. Small groups of women drawn from three generations were asked to discuss the changes that had taken place. They checked, reminded and confirmed each other, enjoyed the discussion and themselves learnt in the process. Similar approaches, with small groups selected to bring together specialised knowledge, may often be an exciting and efficient method of investigation. It is not only the outsider who holds the initiative or who gains; all who take part can influence the direction of the discussion, and be absorbed in learning. Evening meetings may be ideal, going on into the night, when the outsider's presence is less obtrusive and distorted responses less likely.

ii) Learning from the poorest

The poorest are usually considered to be the most ignorant, those from whom there is least to learn. But how much do outsiders know about how the poorest cope? To enable the poorest to do better, the starting point is to understand how they manage at present. And on this the poorest are the experts – they know more than ignorant outsiders who have not bothered to try to find out.

Let one example suffice. Paulus Santosa of the Bethesda Hospital, Jogjakarta, Indonesia, trains nutrition workers through learning from the poorest. He asks, 'Shall we try to teach the under privileged, or shall we rather try to learn about nutrition from them?' Trainee nutrition workers are required to learn from families of the poor whose children are healthy and of normal weight. As Santosa puts it, 'it would be very hard to find professional nutrition workers in Indonesia today who can raise a family of five with US$0.50 per day and stay healthy'. It is after learning how some of the poor manage to have healthy children that plans are formulated with villagers for combatting malnutrition. These are usually in agriculture, animal husbandry, cottage industry, health education, safe water, improved sanitation and the like. They have never included a direct feeding programme, since the trainees have realised, through their research, that it would not solve the problem. Through learning from the poorest, they come to understand that the direct, obvious, solution to physical weakness (see pp. 163–5) is only temporary and a palliative; other measures to tackle poverty itself will be more lasting and effective.

Learning from the poorest is rarely any part of anti-poverty programmes and projects; yet it is a key to enabling them to improve their lot.

iii) Learning indigenous technical knowledge

All rural people know things which outsiders do not know, and some know more than others. There are many ways for outsiders to learn from them. There is the comprehensive approach of a social anthropologist concerned with knowledge systems, including concepts and patterns of thought. But short of this, there are less complicated or abstruse approaches, including compiling glossaries of local terms, and games, quantification and ranking.

a) Glossaries of local terms

Compiling glossaries of local terms and concepts has been proposed by a team working in Nigeria on local participation in environmental monitoring. They suggest

the compilation of a dictionary of local terms and definitions of ecological significance. In some languages, names of plants, insects and soil types are specific to the town or village in question. The identification and systematic recording of local terminology and the investigation of the taxonomic basis of their meanings not only lead to improved communication but often provide information vital to an understanding of local environmental conditions.
(Barker *et al*, 1977, p. 50)

In their own work they found that the literal meanings of names for a harmful insect provided information about its habits and habitat.

The compilation of such glossaries together with descriptions of each item should be easy to incorporate in training courses for rural extension staff in, say, agriculture and health. The range of subjects to be covered depends upon needs, but can include colours, climate, time, soils, plants, topography including micro-environments, animals, insects, foods and diets, diseases, cures, pests, weeds, seasons, space, measurements, proverbs, social relations, and ceremonies. Finding out about these, writing down names, describing and explaining their meanings, and then comparing notes with colleagues, is interesting in its own right. The insight and information elicited is likely to be directly useful. The exercise should improve understanding of local beliefs and practices as well as communication between staff and their public through enabling staff to use categories and expressions which are locally familiar.

b) Games, quantification and ranking

As a way for outsiders to learn from rural people, games have the advantage of suspending status and social differences besides being fun. They can take many forms. To elicit values and constructs social scientists have used sentence completion, and also back-to-back guessing where one respondent describes one of a class of objects (a rice variety, a fertiliser etc.) while another, who cannot see it, guesses the name.

Paul Richards (1979) has used another game to find out the way different groups of people see weeds. Respondents were given different sorts of weeds in groups of three and asked to say which two were most similar and which was most different. They were then asked to explain the 'construct' underlying their choice. This was repeated with different combinations of three weeds.[4] The game, with the same weeds, was given to three groups. The first, Sierra Leonian university botany and geography students, differentiated the weeds by their shape and appearance and by scientific classification. The second group, who were farmers, had totally different, utilitarian, constructs such as ease or difficulty of clearing. The most startling finding was that the third group, extension trainees, had constructs which were almost identical with those of the university students and quite different from those of the farmers. This led to

> a spontaneous 'seminar' by the trainees on how they would communicate with farmers if their 'scientific' approach to farming made them think in textbook botanical terms rather than in terms of farming utilities. Tentative action proposals for syllabus development and for studying alongside the farmers were beginning to emerge at the end of the period.
> (ibid, p. 32)

Without such exercises, outsiders, including agricultural extension staff, may not realise how radically differently they and their clients are seeing and thinking about the same things.

Indigenous methods of counting and quanitification are often supposed to be weak; and this view may be used to justify top-down statistical surveys. Before such surveys are designed and undertaken, however, prior questions can be asked about how rural people measure and count, and whether these methods can obviate or fit into a survey. A first step is to list, investigate and calibrate the units used by rural people and their applications. Farmers' units for land under crops may, for example, represent not area but labour inputs – such as the length of a row that could be worked before standing up and stretching (Richards, 1979, p. 31) – and these units may be more meaningful to farmers and for some survey purposes than standard geometrical measures of area.

Local games can also be used to help farmers quantify and scale their estimates and preferences. David Barker has described how a traditional African board game, *ayo*, involving holes in a board or in the ground and employing seeds or stones, has been adapted to explore farmers' decision-making, their estimates of the relative severity of a pest outbreak, and their preferences for different patterns of farm returns over several years. Farmers in Sierra Leone prefer this game format to a questionnaire survey. The ayo board passes the initiative in providing information to the local people. This 'seems to be very important in oral cultures where questionnaire schedules can act as both steering wheel and brakes on the free flow of discussion' (Barker, 1979, p. 40).

With pastoralists in West Africa, Jeremy Swift has developed a variant of the ayo board to generate discussion about priorities, using camel pellets and holes in the ground. He makes eight holes, and then asks a group to name their eight most important problems, with a hole to represent each. This takes some time and provokes lively debate. He then gives the group 25 camel pellets and asks them to put the pellets in the holes according to the importance of the problem. The lesser problems are gradually eliminated and the pellets redistributed to those that remain. With the two or three biggest problems, the next stage of analysis is to make five further holes for each, and to ask the group to name the most important aspects of each problem. Swift found that the same problems recurred – how to obtain food at reasonable prices in the dry season; how to get credit to build up herds; animal and human health; how to get just treatment in dealings with the Forest Department; how to deal with government and the modern world. One discussion which started at five o'clock in the evening went on until after midnight.

Approaches such as these shift the initiative from the outsider. Things on the ground – drawings on the sand or in the mud, holes, stones or pellets – provide a common focus which distracts from differences of status, dress, style, or speech. The way is open for inventiveness in changing the game or the model through physical acts – taking a stick and drawing on the ground, moving pellets from one hole to another. Vigorous debate reveals a spectrum of views. Teaching by 'respondents' is uninhibited, and learning by outsiders can be deep and direct.

iv) Joint R and D

Other reversals in learning can come from the location and mode of research. The strong reasons for carrying out much agricultural and agricultural engineering research jointly with farmers in their fields and under their conditions are now widely accepted. Research conducted outside the rural environment (on a research station, in a laboratory) often entails heroic simplifications or gross distortions. In the past, much agricultural research undertaken without the small farm and the small farm family has had the wrong priorities and has generated misleading 'findings'. There are, to be sure, some stages or forms of research which require stringent controls or special equipment which only a research station or laboratory can provide. But professional biases weigh heavily towards working in research station and laboratory cores instead of in field condition peripheries.

Here, too, therefore, a reversal is required, recognising small farmers as professionals and colleagues, as fellow experimenters and developers of technology. Peasant farmers in Sierra Leone, for instance, have their own experimental methods which are said to render expensive, supervised on-farm trials unnecessary (Johnny and Richards, 1980). Informal R and D is continuously practised by farmers and there is much to be learnt from it (Biggs, 1980). A handbook for learning about and improving farmers' own experimental methods has yet, to my knowledge, to be written. And the potential of collaboration between farmer experimenters and agricultural scientists is still far from being realised.

v) Learning by working

For many outsiders, there is scope for learning by physically working with farmers and others, and doing what they do. This is not new to social anthropologists. Paul Devitt, living in the remote village of Kuli in the Kalahari desert, kept a dozen cattle in order to understand village life and economy by experiencing it from the inside. Paul Richards found that working at farm tasks with farmers in their fields elicited information he would not have known to ask for and his informants would not have known to volunteer. John Hatch, in Peru, having abandoned a conventional questionnaire survey, worked systematically through the many operations of maize cultivation, hiring himself out to farmers as an unpaid labourer on condition they would teach him the task to be performed

> The scheme worked beautifully. Most small farmers took to their role as teacher very conscientiously. Rather than waiting to respond to my questions, they often volunteered task information I would never have known enough to inquire about. In fact, most of the information I gathered was gained in this way. Hired labourers often proved excellent instructors as well.
>
> (Hatch, 1976, p. 16)

Among other things he found out that crop labour requirements might be half as much again as those previously estimated by outsiders.

These are all examples of individual researchers who found working at rural tasks a source of understanding. But the method has also great possibilities in training. As part of the training of social workers in Madras, Viji Srinivasan required them to buy food for a poor family's meal for only one rupee, and then cook it for the family, assisted only by technical advice from the woman of the family. One lesson brought home to the trainees was how diet is constrained and simplified by having only two cooking pots. The application of similar approaches in training courses would seem to have enormous potential, with foresters collecting firewood or other forest produce, agriculturalists working as farm labourers, animal husbandry staff and veterinarians herding animals, and irrigation engineers and agricultural engineers applying water in farmers' fields; in each case advised by local experts, the rural people themselves.

vi) Simulation games

The most effective way of experiencing the world as a poor person is to go and be that poor person in as complete a manner as possible. But for that most outsiders have neither time, courage nor opportunity. This being so, simulation games are one of the most promising methods for enabling outsiders to understand the life and problems of the poor. Such games are used in other fields such as business management, and war gaming has a long history. Much has been written about them.[5] In development, they have been used in the training of foreign voluntary agency staff and by the Economic Development Institute of the World Bank. Yet in the Third World itself they are little used in training professionals concerned with rural development, such as administrative officers, planners, agricultural extension staff, and those concerned with animal husbandry, education, forestry, health, irrigation and social welfare.

One family of games already developed simulates small farmers. These are the Green Revolution Game (Chapman, 1973; Chapman, Dowler and Elston, 1982), the Peasant Farming Game (Mitchell, 1982), and Ganeshpur (Blaikie, 1982). These are conducted in a single room and participants are usually paired. Each participant or pair is allocated the resources (labour, land, food, money) of a small or near-landless farmer. Decisions are then made, season by season, about crops to grow, inputs to purchase, loans to take, and so on. Random disasters occur and the seasons wait for no one.

The experience of players varies and these simulations take a different form each time. Social relations quickly develop between players (small farm families) as farm employment is sought, money borrowed, and patron-client relations formed. The poorest face a struggle which taxes their energy and ingenuity as they try to save their children from starvation by selling land, by begging for work, loans and food, and even by stealing. The experience of standing in a queue to secure payment for one crop in order to be able to buy inputs for the next, while the rains are imminent and time running out, is not easily forgotten. Nor is the death of a child because an employer was dilatory in paying out wages for work done. To be effective, such simulations need participants who enter into the spirit of the thing, and plenty of time for discussion afterwards to allow participants to remember, describe and analyse what happened and what they felt. Given these, they can be a remarkable way into the experience of others and a powerful source of learning. They can contribute to changes in understanding and in feeling and to a new empathy with poor rural people.

The development and effective use of such games is a priority. They can be applied to many aspects of rural life to capture and illuminate the linkages of poverty with agriculture, the seasons, risk, input supplies, wages, food prices, health, mortality, exploitation, and the position of women. They can enable participants to feel powerlessness and vulnerability, and to experience ratchets of impoverishment. They can show those who provide services what it is like to be poor clients – as bankers play farmers seeking credit, foresters play villagers dependent on forest products, and irrigation engineers play farmers at the tails of canals. Simulations such as these have a key part in training staff and changing bureaucracies. They should be in the curricula of training institutes and of university courses concerned with rural development. To be effective, they must be well designed and well used. They also need to be prepared for different purposes and different environments. The most efficient way to develop them quickly may be workshops for creative researchers, teachers and trainers who will learn how to devise and manage simulation games based on empirical data which they themselves obtain. In the meantime, some good games already exist[6] and should be widely used.

Conclusion

These six approaches – sitting, asking and listening; learning from the poorest; learning indigenous technical knowledge; joint R and D; learning by working; and simulation games – all reverse the learning process. They encourage and enable those being trained or educated to learn from the many below and not just from the few above. They vary in ease and utility. But they share strengths: they transfer initiative to rural people, for them to volunteer information and develop ideas; they encourage an equal relationship between questioner and informant, and the attitude which Peter Berger (1977) calls 'cognitive respect' – respect on the part of the more educated and more influential for the less educated and less influential; and they add to the body of centralised knowledge and understanding. For both rural teachers and outsider students, they can be acts of sharing and learning together.

These and similar approaches should be feasible in many institutions and at many levels. Some have already been tested and are already being used; others are yet to be devised. Researchers and consultants who span the three (academic, practitioner, and rural) cultures are well placed to develop new reversals in learning: a good example is John Hatch's experiment in enabling peasants in Bolivia to write a textbook on their subsistence farming (Hatch, 1981). The adoption of these approaches may, however, face difficulties, notably in those universities and university departments which are fossilised in conservative orthodoxy. It is perhaps in government training institutions that the potential is greatest. Required learning from rural people could help to reorient whole bureaucracies. Nor should it be limited to the junior staff who are most considered to need training. Senior staff should not be deprived of such opportunities for learning. It may even be best to start with the senior staff, both to test and develop the approaches and so that they appreciate the value of the approaches and ensure that they are supported and spread.

Reversals in learning will sometimes be seen as an affront to the dignity, status and professional propriety of outsiders. But learning in reverse can give pride (having worked as a labourer), fun (playing games), the excitement of discovery (learning ITK, how poor people cope, and so on), and practical insight (how realistically to help people to help themselves). It can be intellectually rewarding and practically useful. It can bring together the disciplines. The sociologist and the soil scientist, the agronomist and the economist, can sit down together and find a common activity and common understanding in learning from those other professionals – the rural people – who do not distinguish disciplinary domains. By giving pride, pleasure, intellectual interest, and practical value, reversals in learning ought to appeal to many outsider professionals and gain the wide acceptance they deserve.

Reversals in management

Reversals in learning cannot take place in isolation. They are hindered or supported by the way in which organisations are managed. There are many useful analyses and reviews of management for rural and social development.[7] As usual, local variations of style, culture, structures, procedures and conventions make generalisation difficult, as do the many varieties of organisation involved – government departments, parastatals, banks, voluntary agencies, and business houses. Almost all, though, have a centre-periphery orientation, and those most pointed towards the centre are usually the largest, the field departments of government like agriculture, animal husbandry, community development, education, forestry, health, irrigation, public works and water development.

Most of these departments in most countries face inwards and upwards, away from rural clients. Their structure and style are often authoritarian, hierarchical and punitive. From the centre and top come targets, commands, exhortations and threats. From the periphery and bottom comes a weaker flow of filtered information which placates and misleads. In meetings, subordinates are upbraided, cajoled and given orders. They are asked for reports of targets achieved, not for problems encountered. Poor performance or deviant initiatives are rewarded by punishment or posting. Promotion comes, if at all, through compliance or through working in headquarters. Real problems of implementation or impact are repressed; appearances of achievement applauded. Senior officers do not learn from their subordinates; and subordinates do not learn from their rural clients.

This pattern is common, but neither universal nor inevitable. An exception is the National Irrigation Administration in the Philippines (Alfonso, 1981; Bagadion and F. Korten, 1980; D. Korten, 1980, pp. 492–4) which has evolved a participatory philosophy and style. It is geared to providing services and to responding to its farmer clients. Upward flows of information about problems are encouraged. Social science research is commissioned to explore critical questions and is openly discussed. Staff are allowed to remain for years working in the same post without the disruption of transfer. Learning and improvement within the organisation are a conscious and continuous aim.

The contrast between hierarchical and participatory organisations corresponds with the useful distinction between the blueprint and the learning process approaches (Korten, 1980; Moris, 1981, pp. 19–22). The blueprint approach emphasises planning from the top and then implementation below. The more successful learning process approach starts below. From his analysis of five Asian rural development success stories David Korten concludes that

> the blueprint approach never played more than an incidental role in their development. These five programs were not designed and implemented – rather they emerged out of a learning process in which villagers and program personnel shared their knowledge and resources to create a program which achieved a fit between needs and capacities of the beneficiaries and those of the outsiders who were providing the assistance. Leadership and teamwork, rather than blueprints, were the key elements.
> (Korten, 1980, p. 497)

The learning process approach is easier to achieve for small voluntary agencies than for the great field bureaucracies of government. A major challenge over coming decades is bureaucratic reorientation (Korten and Uphoff, 1981), including a change from authoritarian to participatory styles and a shift in responsiveness from orders from above to demands from below.

For those who work in authoritarian bureaucracies, such change may seem remote. It will not come quickly. The key is to search for 'soft spots' (p. 157), interventions which present the best chances for starting change. The introduction of reversals in learning is one starting point. In addition, three others can be suggested: styles of communication; transfer policy and practice; and enabling and empowering poor clients.

i) *Styles of communication*

Styles of communication can be changed in several ways. Senior staff can alter the tone of their interaction with subordinates. They can talk less and listen more. Procedures can be introduced to require joint planning and programming with junior staff taking part in drawing up their own work plans. Training can support such changes. Target-setting by those above can be reduced or abolished. Efforts can be made to learn from junior staff, and to encourage them to learn from their clients.

Such changes do not come all at once. Nor are individuals always consistent. A District Collector in India is reported to have said

> ... most of my staff have worked under many difficult superiors so they have become more cautious. There is this great willingness to agree. They would rather agree than set out the difficulties and present their objections. They are afraid of rubbing their superior the wrong way. I try to make them feel relaxed and give them a feeling of partnership in these programs. I find this is the best way of doing things. This kind of dialogue leads to them feeling free to express themselves.

Yet earlier that same day his tone to his staff had been one of 'open and rough criticism' (Heginbotham 1975, pp. 126–7) which can hardly have helped them feel free to express themselves. Changes in style to encourage subordinates to be frank and to exercise initiative do not come easily for those trained and socialised in a hierarchical tradition; but they are something that each person in a large bureaucracy, however difficult his or her situation, can work on.

ii) Transfer policies and practice

Rapid transfers are the slipping clutch of rural development administration. Again and again, they incapacitate field organisations of government. Transfers of field staff are made for many reasons, including political pressures, chain reactions in which one posting provokes others, and to solve personnel problems. Of East African administration, Jon Moris has put forward the hypothesis that

> In the face of a lack of enforceable sanctions, the top officials use the transfer of subordinates as the main administrative solution to almost every problem. Consequently, top officials are constantly being moved around the country within the bureaucracy. Problems having to do with corruption or personality are not solved, but merely exported to some other locus in the system.
> (Moris, 1977, p. 79)

Postings seem to be most rapid in the remoter and poorer areas. In East Africa, turnover in district staff in pastoral areas has been so rapid that there could be only very temporary and superficial contact between two systems of nomadism – that of the pastoralists and that of the administrators. In 1970 a District Development Committee meeting in the remote Samburu District in Kenya had as the major item on the agenda a land use plan for part of the district, but only one of the ten or so government officers present had been in the District for more than a year. In India, for the very poor tribal area of Dharampur in Gujarat, Ranjit Gupta (1981, p. 121) reports that in two years there were six different District Collectors, five District Development Officers, and three Project-cum-Tribal Development Officers. He attributes the high rate of transfers to the three Ps: probation for young staff; promotion for some who can only rise by accepting an unpopular posting, after which they at once negotiate for a re-transfer, carrying their promotion with them; and punishment – the penal posting – for those who have misbehaved or otherwise earned disfavour.

It is difficult to exaggerate the bad effects of these practices. Young staff on probation lack the experience to do much. Older staff who are trying to negotiate a re-transfer do not apply their best energies to their work. Staff who are in disgrace may be incompetent, or if competent, have low morale. And staff who expect to be posted in a matter of months have little incentive to get to know the people of an area, or to work for longer-term or participatory development. Reversing this tendency may not be easy; but it is often a precondition for sustained and better support for rural development in poor and remote areas that government staff be able and committed and that they remain in post for a matter of years.

Many measures can be suggested. At the personal level, individual staff can seek to stay in their posts and try to resist transfer; those in senior positions can resist pressures for their subordinates to be moved; politicians can deny themselves short-term gains from moving staff for the sake of longer-term gains from development. At the policy level, rules can be made to require substantial periods of continuous service in poor and remote areas, making this a condition for promotion. But once again popular demand may be the most effective measure. One approach is to publish annually, for each administrative area, the rate of turnover of staff in different departments over the previous few years. In Kenya, the publication of examination results by school and by district led parents in low performance areas to organise, protest, and secure better services and improved performance (Somerset, 1982). The same technique could expose the extent to which poor areas are deprived through rapid transfers of officials and might generate demands for greater stability and better service.

iii) Enabling and empowering poor clients

The third management reversal involves shifting power and initiative downwards and outwards. Decentralisation − the deconcentration of staff and resources, and the devolution of authority − is usually seen as the result of central decisions. To some degree this is inevitable. But there is another angle. Administrative and service organisations are regarded as providers: they spread facilities, pass on knowledge, treat the sick, educate the children, and so on. They can better be seen as enablers, enabling those who are variously poor, powerless and remote to control more of their lives, to have more choice, and to demand and use more services.

For this reversal, two conditions help. The first is clear specification of the rights of poor clients −rights of forest dwellers to forest produce; of small farmers to irrigation water in stated amounts at stated times; of communities to schools and teachers; of mothers and children to health care; of labourers to off-season employment on public works; of poor producers to fair prices for their produce. These rights can be published, proclaimed in meetings, and displayed on noticeboards. The second condition is the formation of groups around common interests, encouraging them to demand their rights – forest dwellers to demand their forest produce and protection of the forest from pillage by contractors; small farmers to demand their irrigation water; communities to demand teachers for their schools; mothers to demand health care; labourers to demand employment; and specialised groups like the fisher-people, charcoal-burners, and honey-gatherers of Zambia, to demand fair prices. The role of the service organisation is, then, not just to provide the service; it is to see that clients know their rights and have power to demand them, enabling them to ensure quality of service and access.

Reversals in management can require vision, courage and effort. Obstacles to them include hierarchy, authoritarianism, corruption, administrative inertia, and fear of fostering popular demands. But there are enough positive examples to give grounds for hope. Some are reported in the substantial literature on participation.[8] Many, many others represent the quiet personal achievements of people determined, in however small a way, to make things better than they would have been.

The primacy of personal action

So it comes back to personal choice. Analyses of 'the problems of rural development' again and again halt at the stage of a general statement which suggests that nothing can be done unless something else is done first. Once participatory management has been introduced, or corruption wiped out, or training reorganised, or personnel policies reformed, or political interference reduced, or a new head of department appointed, or the workload rationalised, or target-setting abolished, or whatever, then and only then, it is tempting to say, will better performance be possible. But these are excuses. There is always *something* that can be done; *some* scope for personal action. Anyone can ask these questions:

how do my actions affect the poorer rural people?
how should I and can I change them?
what more can I do?
how can I help others to do more?

Some may feel that their position is hopeless, that there is nothing they can do. The 'system' is too strong for them. Perhaps the best antidote to this despair is to study the examples and lives of those who have fought against the odds and succeeded. In every country there are some courageous people – political and religious leaders, civil servants, workers in voluntary agencies, academics, scientists, and others – who have refused to give in, who have stuck by their principles and whose lives shine as examples to others of what can be done. For those who side with the poor, too, there may be unexpected floods of support. Arun Bhatia, the Indian District Collector who exposed corruption in the Maharashtra Employment Guarantee Scheme, received massive popular acclaim for his actions, even though in the short term he was transferred to another post.[9] But not all can expect recognition or to become folk-heroes. For most of those who put the last first, the satisfaction and rewards are not fame, but in knowing that they have done what was right, and that things are, however slightly, better than they would have been. Their small deeds may not command attention; but in merit, they may equal or exceed the greater and more conspicuous actions of those with more freedom and power.

Personal choices are both strategic and tactical. At the strategic level, there are choices about career and specialisation. More professionals are needed who span the three cultures and who organise their lives so that they move more freely between them. More are needed, too, who are multi-disciplinary not just in tolerance but in themselves. Many who make the greatest contributions are those who decide to spend more of their lives working in difficult conditions in difficult places.

At the tactical level there are choices of timing and degree. Not everything can be done at once. Small gains well consolidated as part of a sequence can mean more than big gains which are unstable and short-lived. There are times for confrontation and the big reversal; there are critical periods when small pushes can move major decisions, resources or systems one way or another; but most common are the times for patient work on small things: the items on the agenda paper and their order (is the last first?); the question raised and the argument put in the committee; the condition negotiated in the agreement; the detail of the syllabus in the training course; the programme for the VIP; the way visitors behave and the questions they ask. Small moves to put the last first all count, and they add up.

It is action that matters. Much of the analysis in this book has been about knowing – about how outsiders perceive or do not perceive rural deprivation. But knowing does not guarantee a change of feeling; and a change of feeling does not guarantee a change of behaviour. So we come to the final, paradoxical, reversal: to start by acting. Changes in feeling and perception can come back to front, from changed behaviour and the experiences it generates. The traditions of science, scholarship and management are to begin with data collection, analysis, and planning, often protracted, often delaying action. But there are usually some obvious things that can be done at once. Not everything can or should be foreseen. It is often best to start, to do something, and to learn from doing.

For the test is what people do. Social change flows from individual actions. By changing what they do, people move societies in new directions and themselves change. Big simple solutions are tempting but full of risks. For most outsiders, most of the time, the soundest and best way forward is through innumerable small steps and tiny pushes, putting the last first not once but again and again and again. Many small reversals then support each other and together build up towards a greater movement. The lives of many people already show a will to make reversals, to put the last first. Some contribute from a distance. Others work directly with and for those who are rural and poor, helping them to gain more of what they want and need and to demand and control more of the benefits of development. The hope of this book is to support all those people, in however slight a way, and to encourage others to do what they can.

Notes

1 See Barnett, 1979, for a review of a workshop on RRA. The rationale and some techniques of RRA are described in Agricultural Administration, 1981; Chambers, 1981; Longhurst, 1981a; and Pacey, 1981; each of which lists references. Rhoades, 1982, is a practical guide to the informal agricultural survey. The Oxfam Field Directors' Handbook (Oxfam, 1980) is an admirable source of information and guidance about rural conditions and development.

2 There is a large literature. See Gilbert, Norman and Winch, 1980, for a selected bibliography, and the reports of the Farming Systems Group at Michigan State University.

3 Four sources are James and Jongeward *Born to Win: Transactional Analysis with Gestalt Experiments*, 1971; Fagan and Shepherd, *Gestalt Therapy Now*, 1972; Binder, Binder and Rimland, *Modern Therapies*, 1976; and Rowan, *Ordinary Ecstasy: Humanistic Psychology in Action*, 1976.

4 The method is called the triads test and is connected with G. A.

Kelly's personal construct theory. See Kelly, 1955, and Fransella and Bannister, 1977.

5   As a start, see Oxenham, 1982, which contains short descriptions of several games, and overview articles by Oxenham and Walford, who also provides a short bibliography.

6   The Green Revolution Game is available on sale from Marginal Context Limited, 36 St. Andrew's Road, Cambridge CB4 1DL, UK.

7   A short list is Lele, 1975; Leonard, 1977; Hunter, 1978; Bryant and White, 1980; Knight, 1980 (especially for Esman and Montgomery;) Korten, 1980; Korten and Alfonso, 1981; Moris, 1981.

8   For a review see Cohen and Uphoff, 1977. Good sources are the *Rural Development Participation Review* and other publications of the Rural Development Committee, Cornell University, Ithaca, New York.

9   *Indian Express*, 22 June and 29 July 1982, and *Times of India*, 18 September 1982.

References

Adams, Adrian, 1979, 'An Open Letter to a Young Researcher', *African Affairs*, 78, No. 313, October, pp. 451–79

Agarwal, Bina, 1980, *The Woodfuel Problem and the Diffusion of Rural Innovations*, a report to the Tropical Products Institute, UK; October.

Agricultural Administration, 1981, *Special Issue – Rapid Rural Appraisal*, 8, 6, edited by Ian Carruthers and Robert Chambers, Applied Science Publishers, Barking, Essex, UK; November.

Alfonso, Felipe B., 1981, 'Assisting Farmer Controlled Development of Communal Irrigation Systems', in Korten and Alfonso (eds), *Bureaucracy and the Poor*, pp. 44–52.

Allan, William, 1965, *The African Husbandman*, Oliver and Boyd, Edinburgh and London.

Arens, Jenneke and van Beurden, Jos, 1977, *Jhagrapur: Poor Peasants and Women in a village in Bangladesh*, Third World Publications, 151 Stratford Road, Birmingham, UK.

Arole, R. S., 1977, 'Community Action – Family Health Programmes Delivering an Integrated Package. Illustrative Case Study: India', paper for the IUNS Working Conference held at the National Institute of Nutrition, Hyderabad; 17–21 October.

Ashe, Jeffrey, 1979, *Assessing Rural Needs: A Manual for Practitioners*, VITA (Volunteers in Technical Assistance), 3706 Rhode Island Avenue, Mt Rainier, Maryland 20822, USA.

Bagadion, Benjamin Jr., Espiritu, Augusto Cesar, *et al.*, 1979, *Law in the Mobilization and Participatory Organization of the Rural Poor: The Kagawasan Case*, Institute of Philippine Culture, Ateneo de Manila University, Quezon City, Philippines.

Bagadion, Benjamin U., and Korten, Frances F., 1980, 'Developing Viable Irrigators' Associations: Lessons from Small Scale Irrigation in the Philippines', *Agricultural Administration*, 7, pp. 273–87.

Bailey, F. G., 1957, *Caste and the Economic Frontier: a Village in Highland Orissa*, Manchester University Press, Manchester, UK.

Banfield, E. C., 1958, *The Moral Basis of Backward Society*, The Free Press, New York.

Barker, David, Oguntoyinbo, Julius, and Richards, Paul, 1977, *The Utility of the Nigerian Peasant Farmer's Knowledge in the Monitoring of Agricultural Resources*, MARC Report Number 4, Monitoring and Assessment Research Centre, Chelsea College, University of London, London.

Barker, David, 1979, 'Appropriate Methodology: An Example using a Traditional African Board Game to measure Farmers' Attitudes and Environmental Images', *IDS Bulletin* 10, 2, pp. 37–40. Also Brokensha

References

et al., 1980, pp. 301–6.

Barnett, Andrew, 1979, 'Rapid Rural Appraisal: a personal view of the first IDS workshop', paper for the second workshop on Rapid Rural Appraisal, Institute of Development Studies, University of Sussex, Brighton; 4–7 December.

Belshaw, D. G. R., and Hall, M., 1972, 'The Analysis and Use of Agricultural Experimental Data in Tropical Africa', *East African Journal of Rural Development*, No. 5.

Belshaw, Deryke, 1979, 'Taking Indigenous Technology Seriously: the case of intercropping techniques in East Africa', *IDS Bulletin*, 10, 2, pp. 24–7. Also in Brokensha et al., 1980, pp. 197–203.

Berger, Peter L., 1977, *Pyramids of Sacrifice: Political Ethics and Social Change*, Penguin Books (first published 1974 by Basic Books), Harmondsworth, UK.

Berlin, Isaiah, 1953, *The Hedgehog and the Fox: an essay on Tolstoy's view of history*, Weidenfeld and Nicholson, London.

Biggs, Stephen, 1980, 'Informal R and D', *Ceres*, 76, Vol. 13(4), July–August, pp. 23–6.

Binder, V., Binder, A., and Rimland, B. (eds), 1976, *Modern Therapies*, Prentice-Hall Inc., Englewood Cliffs, New Jersey.

Blaikie, M. P., Cameron, J., and Seddon, J. D., 1979, *The Struggle for Basic Needs in Nepal*, Development Centre Studies, Development Centre of the Organisation for Economic Cooperation and Development, Paris.

Blaikie, P. M., 1972, 'Implications of Selective Feedback in Aspects of Family Planning Research for Policy-Makers in India', *Population Studies*, 26, 3 November, pp. 437–44.

Blaikie, Piers, 1982, 'Ganeshpur', in John Oxenham (ed.), *Simulations and Adult Learning for Development*, pp. 25–7.

Blurton Jones, Nicholas, and Konner, Melvin J., 1976. '!kung Knowledge of Animal Behaviour (or: The Proper Study of Mankind is Animals)', in Lee and deVore (eds), *Kalahari Hunter-Gatherers*, pp. 325–48.

Bornstein, M. H., 1975, 'The Influence of Visual Perception on Culture', *American Anthropologist*, 77, pp. 774–98. Cited in Thomasson, 1979, pp. 170–1.

Boserup, Ester, 1965, *The Conditions of Agricultural Growth: the Economics of Agrarian Change under Population Pressure*, George Allen and Unwin, London.

Bottrall, Anthony, 1981, *Comparative Study of the Management and Organization of Irrigation Projects*, World Bank Staff Working Paper No. 458, World Bank, 1818 – Street NW, Washington DC; May.

Boxall, R. A., Greeley, M., Tyagi, D. S., Lipton, M., and Neelakanta, J., 1978, 'The Prevention of Farm-Level Foodgrain in Storage Losses in India: A Social Cost-Benefit Analysis', IDS Research Report, Institute of Development Studies, University of Sussex, Brighton, UK; October.

Bowen, Elenore Smith, 1965, *Return to Laughter*, Natural History Press, Doubleday, New York.

BRAC, 1979, *Peasant Perceptions: Famine*, Bangladesh Rural Advancement Committee, 66 Mohakhali Commerical Area, Dacca 12, Bangladesh; July.

BRAC, 1980, *The Net: Power Structure in Ten Villages*, Bangladesh Rural Advancement Committee, 66 Mohakhali Commercial Area, Dacca 12, Bangladesh; February.

Brammer, Hugh, 1977, Personal communication cited in G. W. Olson, 1977, 'Training Key People in Soil Survey Interpretations in Southeast Asia', Agronomy mimeo 77–15, Department of Agronomy, Cornell University, Ithaca. Cited in Thomasson 1978, p. 172.

Brammer, Hugh, 1980, 'Some Innovations Don't Wait for Experts', *Ceres*, 132, March–April, pp. 24–8.

Breman, Jan, 1979, *Patronage and Exploitation: Changing Agrarian Relations in South Gujarat*, Manohar Publications, New Delhi (first published 1974).

Brokensha, David W., Warren, D. M., and Werner, Oswald (eds), 1980, *Indigenous Systems of Knowledge and Development*, University Press of America Inc., Lanham, MD 20801, USA.

Brokensha, David and Riley, Bernard W. 1980, 'Mbeere Knowledge of their Vegetation and its Relevance for Development: a Case-study from Kenya', in Brokensha *et al.* (eds), pp. 113–29.

Brush, Stephen, 1980, 'Potato Taxonomies in Andeen Agriculture' in Brokensha *et al.* (eds).

Bryant, C. and White, L. 1980, *Managing Rural Development: Peasant Participation in Rural Development*, Kumarian Press, West Hartford, Connecticut.

Byres, T. J., 1972, 'The Dialectic of India's Green Revolution', *South Asian Review*, 5, 2, January, pp. 99–116.

Cain, Mead, 1981, 'Risk and Insurance: Perspectives on Fertility and Agrarian Change in India and Bangladesh', *Population and Development Review*, 7, 3, September.

Campbell, J. Gabriel, Shrestha, Ramesh, and Stone, Linda, 1979, *The Use and Misuse of Social Science Research in Nepal*, Research Centre for Nepal and Asian Studies, Tribhuvan University, Kirtipur, Kathmandu.

Cassen, Robert H., 1976, 'Population and Development: a Survey', *World Development*, 4, 10/11.

CEC, 1977, *The Perception of Poverty in Europe*, Commission of the European Communities, Rue de la Loi, 1049 Brussels; March.

Chambers, Robert, 1969, 'Report on Social and Administrative Aspects of Range Management Development in the North Eastern Provinces of Kenya', mimeo, Ministry of Agriculture, Nairobi.

Chambers, Robert, 1974, *Managing Rural Development: Ideas and Experience from East Africa*, Scandinavian Institute of African Studies, Uppsala.

Chambers, Robert, 1979, 'Health, Agriculture and Rural Poverty: Why Seasons Matter', *IDS Discussion Paper*, 148, Institute of Development Studies, University of Sussex, Brighton, UK; December.

Chambers, Robert, 1979, 'Rural Refugees in Africa: What the Eye Does Not See', *Disasters*, Vol. 3, No. 4.

Chambers, Robert, 1980, 'In Search of a Water Revolution: questions for managing canal irrigation in the 1980s', in *Irrigation Water*

References

Management, Report of a Planning Workshop, International Rice
Research Institute, Los Banos, Laguna, Philippines, pp. 23–37.

Chambers, Robert, 1981, 'Rapid Rural Appraisal: Rationale and
Repertoire', *Public Administration and Development*, 1, 2,
pp. 95–106. (Also Discussion Paper 155, Institute of Development
Studies, University of Sussex, Brighton, UK.)

Chambers, Robert, Longhurst, Richard, and Pacey, Arnold (eds), 1981,
Seasonal Dimensions to Rural Poverty, Frances Pinter, London.

Chambers, Robert, 1982, 'Health, Agriculture and Rural Poverty: Why
Seasons Matter', *Journal of Development Studies*, 18, 2, January.

Chapman, G. P., 1973, 'The Green Revolution: a gaming simulation',
Area, Institute of British Geographers, 5, 2, pp. 129–40.

Chapman, G.P., 1977, 'The Folklore of the Perceived Environment in
Bihar', mimeo, Department of Geography, Cambridge University, UK.

Chapman, G. P., Dowler, E., and Elston, M., 1982, 'Experiencing the
Green Revolution', in Oxenham (ed.), *Simulations and Adult Learning
for Development*, pp. 20–24.

Chinnappa, B. Nanjamma, 1977, 'Adoption of the New Technology in
North Arcot District', in Farmer, B. H. (ed.), *Green Revolution?*
Macmillan, London and Basingstoke, pp. 92–123.

Chinnappa, B. Nanjamma, 1977, 'The North Arcot Sample Survey', in
Farmer, B. H. (ed.), *Green Revolution?* Chapter 5, pp. 37–44.

Chuta, Enyinna and Liedholm, Carl, 1979, 'Rural Non-Farm
Employment: A Review of the State of the Art', MSU Rural
Development Paper No. 4, Department of Agricultural Economics,
Michigan State University, East Lansing, Michigan 48824.

CIMMYT, 1977a, *Demonstrations of a Interdisciplinary Farming
Systems Approach to Planning Adaptive Agricultural Research
Programmes, Report No. 1, April 1977: Part of Siaya District, Nyanza
Province, Kenya*, Cimmyt Eastern African Economics Programme, P.O.
Box 25171, Nairobi.

CIMMYT, 1977b, as above, *Report No. 2 December 1977. The Drier Areas
of Morogoro and Kilosa Districts, Tanzania.*

CIMMYT, 1978, as above, *Report No. 3, Part of Serenje District, Central
Province, Zambia.*

CIMMYT, 1980, *Planning Technologies Appropriate to Farmers:
Concepts and Procedures*, Economics Program, CIMMYT, Apdo.
Postal 6-641, Mexico 6, D. F. Mexico.

CIMMYT Economics Staff, 1981, 'Assessing Farmers' Needs in
Designing Agricultural Technology', IADS Occasional Paper,
International Agricultural Development Service, 1133 Avenue of the
Americas, New York, NY 10036.

Cobb, Richard, Hunt, Robert, Vandervoort, Charles, Bledsoe, Caroline,
and McClusky, Robert, 1980, *Impact of Rural Roads in Liberia*, Project
Impact Evaluation No. 6, June, Agency for International Development,
Washington.

Cohen, John M. and Uphoff, Norman, 1977, 'Rural Development
Participation: Concepts and Measures for Project Design,
Implementation and Evaluation', Rural Development Monograph No.

2, Center for International Studies, Cornell University, Ithaca, New York.

Collinson, Michael, 1981, 'A Low Cost Approach to Understanding Small Farmers', *Agricultural Administration*, 8, 6, November, pp. 433–50.

Conklin, Harold C, 'An Ethnoecological Approach to Shifting Agriculture', in Andrew P. Vayda (ed.), *Environmental and Cultural Behaviour: ecological studies in cultural anthropology*, The Natural History Press, Garden City, New York, pp. 221–33. Cited in Howes, 1979, p. 13.

Conlin, Sean, 1979, 'Baseline Surveys: an escape from thinking about research problems and even more a refuge from actually doing anything', paper to the Conference on Rapid Rural Appraisal, IDS, University of Sussex, Brighton, UK; 4–7 December.

Dalby, D., 1964, 'The noun gàrii in Hausa: a semantic study', *Journal of African Languages*, 3, pp. 273–305. Cited in Langley, 1975, p. 93.

Dandekar, V. M., and Rath, Nilakantha, 1971, *Poverty in India*, Indian School of Political Economy, reprinted from *Economic and Political Weekly*, Bombay, 6, Nos. 1 and 2, 2 and 9 January.

Darling, Sir Malcolm, 1947, *The Punjab Peasant in Prosperity and Debt*, Geoffrey Cumberlege, Oxford University Press, London (first published 1925, fourth edition 1947).

Dasgupta, Biplab, 1975, 'A Typology of Village Socio-economic Systems from Indian Village Studies', *Economic and Political Weekly*, Vol. X, Nos. 33–35, pp. 1394–414, August.

Dasgupta, Biplab, 1977, *Agrarian Change and the New Technology in India*, United Nations Research Institute for Social Development, Geneva.

de Schlippe, Pierre, 1956, *Shifting Cultivation in Africa: the Zande System of Agriculture*, Routledge and Kegan Paul, London.

Devitt, Paul, 1977, 'Notes on Poverty-orientated Rural Development', in *Extension, Planning and the Poor*, Agricultural Administration Unit Occasional Paper 2, Overseas Development Institute, 10–11 Percy Street, London W1P 0JB, pp. 20–41.

DF, 1979, *Development Forum*, June/July, p. 14.

Dias, H. D., 1977, 'Selective Adoption as a Strategy for Agricultural Development: lessons from Adoption in S.E. Sri Lanka', in Farmer (ed.), *Green Revolution?*, pp. 54–84.

Dolci, Danilo, 1966, *Poverty in Sicily, a Study of the Province of Palermo* (1956), Penguin Books, Harmondsworth, UK.

Early, Alan C., 1980, 'An Approach to Solving Irrigation System Management Problems', in *Irrigation Water Management*, Report of a Planning Workshop, International Rice Research Institute, Los Banos, Laguna, Philippines, pp. 83–113.

Eckholm, Erik P., 1976, *Losing Ground: Environmental Stress and World Food Prospects*, Pergamon Press, Oxford, UK.

References

Ehrlich, Paul R., 1968, *The Population Bomb*, 1971 edition, Pan Books, London.

Elliott, C. M., 1970, 'Effects of Ill-health on Agricultural Productivity in Zambia', in A. H. Bunting (ed.), *Change in Agriculture*, Duckworth, London, pp. 647–55.

Elliott, Charles, 1975, *Patterns of Poverty in the Third World: A Study of Social and Economic Stratification*, Praeger, New York, Washington, London.

Elliott, Charles, 1982, 'The Political Economy of Sewage: a Case Study from the Himalayas', *Mazingira.*, 6, 4, pp. 44–56

Epstein, Scarlett, 1973, *South India: Yesterday, Today and Tomorrow*, Macmillan, London and Basingstoke.

Esman, Milton J., 1978, *Landlessness and Near-Landlessness in Developing Countries*, Rural Development Committee, Center for International Studies, 170 Uris Hall, Cornell University, Ithaca, NY 14853; September.

Esman, Milton J., and Montgomery, John D., 1980, 'The Administration of Human Development', in Peter T. Knight (ed.), *Implementing Programs of Human Development*, pp. 183–234.

Fagan, Joen, and Shepherd, Irma Lee (eds), 1972, *Gestalt Therapy Now*, Penguin Books, Harmondsworth, UK.

Farmer, B. H. (ed.), 1977, *Green Revolution? Technology and Change in Rice-Growing Areas of Tamil Nadu and Sri Lanka*, Macmillan, London and Basingstoke.

Fitzgerald, Mary Anne, 1980, 'Kenyan tea: an industry in leaf', *Development Forum*, July–August, p. 16.

Fonaroff, Arlene, 1975, 'Cultural Perceptions and Nutritional Disorders: a Jamaican Case Study', *Bulletin of the Pan American Health Organisation*, 9, 3, pp. 112–23.

Frankel, F., 1971, *India's Green Revolution: Economic Gains and Political Costs*, The University Press, Princeton, USA.

Fransella, F., and Bannister, D., 1977, *A Manual for Repertory Grid Techniques*, Academic Press, London.

Freire, Paulo, 1970, *Pedagogy of the Oppressed*, The Seabury Press, New York.

French, M. H., 1970, *Observations on the Goat*, FAO Animal Production and Health Series No. 14, Food and Agriculture Organisation, Rome.

Galbraith, John Kenneth, 1979, *The Nature of Mass Poverty*, Harvard University Press, Cambridge, Mass., and London, UK.

Ganewatta, P., 1974, 'Socio-economic Factors in Rural Indebtedness', *Occasional Publication Series*, No. 7, Agrarian Research and Training Institute, Colombo; February.

George, Susan, 1976, *How the Other Half Dies: The Real Reasons for World Hunger*, Penguin Books, Harmondsworth, UK.

Gilbert, E. H., Norman, D. W., and Winch, F. E., 1980, 'Farming Systems Research: a Critical Appraisal', MSU Rural Development Paper No. 6,

Department of Agricultural Economics, Michigan State University, East Lansing, Michigan.

Goering, T. James, 1979, 'Tropical Root Crops and Rural Development', Work Bank Staff Working Paper No. 324, April, The World Bank, 1818 H Street NW, Washington DC 20433.

Gordon, Gill, 1979, 'Finding Out About Child (0–5 Years) Feeding Practices', paper to the Workshop on Rapid Rural Appraisal, Institute of Development Studies, University of Sussex, Brighton, UK; 4–7 December.

Gore, Charles, 1978, 'The Terms of Trade of Food Producers as a Mechanism of Rural Differentiation', *IDS Bulletin*, 9, 3.

Gosselin, G., 1970, *Le changement social et les institutions du développement dans une population refugiée*, United Nations Research Institute for Social Development and Office of the United Nations High Commissioner for Refugees, Geneva.

Grace, M. R. 1977, *Cassava Processing*, FAO Plant Production and Protection Series No. 3, FAO, Rome.

Greeley, Martin, 1982, 'Pinpointing post harvest food losses', *Ceres*, 15, 1, No. 85, January/February, pp. 30–37.

Griffin, Keith and Khan, Azizur Rahman, 1978, 'Poverty in the Third World: Ugly Facts and Fancy Models', *World Development*, 6, 3, pp. 295–304.

Gulati, Leela, 1981, *Profiles in Female Poverty, a study of five poor working women in Kerala*, Hindustan Publishing Corporation (India), Delhi 110007.

Gupta, Ranjit, 1981, 'The Poverty Trap: Lessons from Dharampur', in Korten and Alfonso (eds), *Bureaucracy and the Poor*, pp. 114–35.

Hameed, N. D. Abdul, *et al.*, 1977, *Rice Revolution in Sri Lanka*, United Nations Research Institute for Social Development, Geneva.

Haque, Wahidul, Mehta, Niranjan, Rahman, Anisur, and Wignaraja, Ponna, 1977, *An Approach to Micro-level Development: Designing and Evaluation of Rural Development Projects*, United Nations Asian Development Institute; March.

Harrison, Paul, 1979a, *Inside the Third World*, Pelican Books, Harmondsworth, UK.

Harrison, Paul, 1979b, 'The Curse of the Tropics', *New Scientist*, 22 November, pp. 602–03.

Harriss, Barbara, 1977, 'Paddy-Milling: Problems in Policy and Choice of Technology', in Farmer (ed.), *Green Revolution?*, pp. 276–300.

Harriss, John, 1977, 'Bias in Perception of Agrarian Change in India', in Farmer (ed.), *Green Revolution?*, pp. 30–36.

Harriss, John, 1982, *Capitalism and Peasant Farming: agrarian structure and ideology in Northern Tamil Nadu*, Oxford University Press, Bombay, Delhi, Calcutta, Madras.

Harwood, Richard R., 1979, *Small Farm Development: Understanding and Improving Farming Systems in the Humid Tropics*, Westview Press, Boulder, Colorado, USA.

Haswell, Margaret, 1975, *The Nature of Poverty: a case-history of the first* quarter-century after World War II, Macmillan, London and Basingstoke.

Hatch, John K., 1976, *The Corn Farmers of Motupe: a Study of Traditional Farming Practices in Northern Coastal Peru*, Land Tenure Center, Monographs No. 1. Land Tenure Center, 310 King Hall, 1525 Observatory Drive, University of Winsconsin-Madison, Madison, Wisconsin 53706.

Hatch, John, 1981, 'Peasants Who Write a Textbook on Subsistence Farming: Report on the Bolivian Tradition Practices Project', *Rural Development Participation Review*, 2, 2, Rural Development Committee, Cornell University, Ithaca, New York, Winter, pp. 17–20.

Hayami, Yujiro, 1978, *Anatomy of a Peasant Economy: a Rice Village in the Philippines*, The International Rice Research Institute, Los Banos, Laguna, Philippines.

Henderson, Willie, 1980, 'Letlhakeng: a Study of Accumulation in a Kalahari Village', Ph.D. thesis, University of Sussex, Brighton, UK.

Heginbotham, Stanley J., 1975, *Cultures in Conflict: the Four Faces of Indian Bureaucracy*, Columbia University Press, New York and London.

Heinz, H. J., and Maguire, B., n.d., 'The Ethno-Biology of the !ko Bushmen: their ethno-botanical knowledge and plant lore', Occasional Paper No. 1, Botswana Society, printed by the Government Printer, Gaborone.

Heyer, Judith, 1981, 'Rural Development Programmes and Impoverishment: Some Experiences in Tropical Africa', in G. Johnson and A. Maunder (eds), *Rural Change, The Challenge for Agricultural Economists*, Proceedings, Seventeenth International Conference of Agricultural Economists, Gower Publishing Co. Ltd., Westmead, Farnborough, Hants, UK, pp. 215–25.

Hildebrand, Peter E., 1981, 'Combining Disciplines in Rapid Appraisal: the Sondeo Approach', *Agricultural Administration*, 8, 6, November, pp. 423–32.

Hill, Polly, 1972, *Rural Hausa: a village and a setting*, Cambridge University Press, Cambridge, UK.

Hill, Polly, 1977, *Population, prosperity and poverty: Rural Kano 1900 and 1970*, Cambridge University Press, Cambridge, UK.

Hirschman, Albert, 1967, *Development Projects Observed*, The Brookings Institution, Washington, DC.

Hirschman, Albert, 1970, *Exit, Voice and Loyalty*, Harvard University Press, Cambridge, Mass.

Hunter, Guy, 1969, *Modernizing Peasant Societies: a comparative study in Asia and Africa*, Oxford University Press, London.

Hunter, Guy, 1970, *The Administration of Agricultural Development: Lessons from India*, Oxford University Press, London.

Hunter, Guy, 1978, *Agricultural Development and the Rural Poor*, Overseas Development Institute, 10 Percy Street, London W1P 0JB.

Hoskins, Marilyn W., 1979, 'Community Participation in African Fuelwood Production, Transformation and Utilisation', Discussion Paper prepared for Workshop on Fuelwood and Other Renewable

Fuels in Africa, Paris, Overseas Development Council, AID, 29–30 November. Quoted in Agarwal, 1980, p. 112.

Howes, Michael, 1979, 'The Uses of Indigenous Technical Knowledge in Development', *IDS Bulletin*, 10, 2, pp. 12–13. Also in Brokensha *et al.*, 1980, pp. 341–57.

Howes, Michael and Robert Chambers, 1979, 'Indigenous Technical Knowledge: Analysis, Implications and Issues', *IDS Bulletin*, 10, 2, pp. 5–11. Also in Brokensha *et al.*, 1980, pp. 329–40.

Howes, Michael, 1980, 'A Year in the Life of a Poor Farming Household', background paper for Inner London Education Authority Schools Project, mimeograph, Institute of Development Studies, University of Sussex, Brighton, UK.

Hyden, Goran, 1980, *Beyond Ujamaa in Tanzania: Underdevelopment and an Uncaptured Peasantry*, Heinemann, London.

IADS, 1981, *Assessing Farmers' Needs in Designing Agricultural Technology*, IADS Occasional Papers, International Agricultural Development Service, 1133 Avenue of the Americas, New York, 10036.

IDS, 1979, 'Rural Development: Whose Knowledge Counts?', *IDS Bulletin*, 10, 2, January, Institute of Development Studies, University of Sussex, Brighton, UK.

IDS, 1981, 'Rapid Rural Appraisal: Social Structure and Rural Economy', *IDS Bulletin*, 12, 4, edited by Richard Longhurst, Institute of Development Studies, University of Sussex, Brighton, UK.

ILO, 1977, *Poverty and Landlessness in Rural Asia*, International Labour Office, Geneva.

ILO, 1981, *Zambia: Basic Needs in an Economy under Pressure*, International Labour Office, Jobs and Skills Programme for Africa, Addis Ababa (available from ILO Publications, ILO, CH-1211, Geneva 22).

James, Muriel, and Jongeward, Dorothy, 1971, *Born to Win: Transactional Analysis with Gestalt Experiments*, Addison-Wesley Publishing Company, Reading, Massachusetts.

Jansen, William H., 1978, December, *Profiles of Poverty in Bangladesh: a preliminary report*, Program Office, USAID Mission to Bangladesh.

Jayaraman, P., and Jayaraman, T. K., 1981, 'Attitudes of the Irrigation Bureaucracy in India to Scientific Water Management Tasks in Irrigated Agriculture: a case study from Gujarat State', *Zeitschrift für Auslandische Landwirtschaft*, Quarterly Journal of Agriculture, Berlin Technical University.

Jodha, N. S., 1978, 'Effectiveness of Farmers' Adjustments to Risk', *Economic and Political Weekly, Review of Agriculture*, June, pp. A38–A48.

Johnny, M. M. P., and Richards, P., 1980, 'Playing with facts: the articulation of "alternative" viewpoints in rural development', Conference on Folk Media in Development, Berlin (Proceedings edited by N. Coletta and R. Kidd, forthcoming).

References

Kannan, K. P., 1981, 'A People's Science Movement', *Development*, 1981:1, pp. 37–40.

Kanwar, J. S., 1982, 'Managing Soil Resources to Meet the Challenges to Mankind', Presidential Address, 12th International Congress of Soil Science, New Delhi; 8–16 February.

Kearl, Bryant (ed.), 1976, *Field Data Collection in the Social Sciences: Experiences in Africa and the Middle East*, Agricultural Development Council Inc., 1290 Avenue of the Americas, New York, NY 10019.

Kelly, G. A., 1955, *The Psychology of Personal Constructs*, Norton, New York.

Knight, Peter T., 1980, 'Implementing Programs of Human Development', World Bank Staff Working Paper No. 403, July, World Bank, 1818 H Street NW, Washington, D.C. 20433.

Korten, David C., 1980, 'Community Organization and Rural Development: a Learning Process Approach', *Public Administration Review*, September/October, pp. 480–511. (An abbreviated version appears in *Rural Development Participation Review*, 2, 2, Winter 1981, Rural Development Committee, Cornell University, Ithaca, New York, pp. 1–8.)

Korten, David C., and Alfonso, Felipe B., 1981, *Bureaucracy and the Poor: Closing the Gap*, McGraw-Hill International Book Company, Singapore and elsewhere.

Korten, David C., and Uphoff, Norman, 1981, 'Bureaucratic Reorientation for Participatory Rural Development', *NASPAA Working Papers*, National Association of Schools of Public Affairs and Administration, Washington.

Korten, Frances F., 1981, 'Stimulating Community Participation: Obstacles and Options at Agency, Community and Social Levels', *Rural Development Participation Review*, 2, 3, Winter, Rural Development Committee, Cornell University, Ithaca, New York, pp. 1–6.

Kurien, C. T., 1978, *Poverty, Planning and Social Transformation*, Allied Publishers Private Limited, Bombay, Calcutta, Madras, New Delhi, Bangalore.

Ladejinsky, Wolf, 1969a, 'The Green Revolution in Punjab: A Field Trip', *Economic and Political Weekly*, 4, 26, 28 June.

Ladejinsky, Wolf, 1969b, 'The Green Revolution in Bihar – the Kosi Area: a Field Trip', *Economic and Political Weekly*, 4, 39, 27 September.

Langley, Philip, 1975, 'The Ethnolinguistic Approach to the Rural Environment: its Usefulness in Rural Planning in Africa', in Paul Richards (ed.), *African Environment Special Report 1, Problems and Perspectives*, International Africa Institute, London, pp. 89–97.

Ledesma, Antonio J., 1977, 'The Sumagaysay Family: a Case Study of Landless Rural Workers', *Land Tenure Center Newsletter*, No. 55, January-March, Land Tenure Center, University of Wisconsin.

Lee, Richard B., and De Vore, Irven (eds), 1976, *Kalahari Hunter-Gatherers: studies of the !kung San and their neighbours*, Harvard University Press, Cambridge, Mass., and London, UK.

Lele, Uma, 1975, *The Design of Rural Development: Lessons from Africa*, The Johns Hopkins University Press, Baltimore and London.

Leonard, David K., 1977, *Reaching the Peasant Farmer: Organization Theory and Practice in Kenya*, University of Chicago Press, Chicago and London.

Lewis, Oscar, 1959, *Five Families: Mexican Case Studies in the Culture of Poverty*, Basic Books, New York.

Leys, Colin, 1975, *Underdevelopment in Kenya: The Political Economy of Neo-Colonialism, 1964–1971*, Heinemann, London.

Lipton, Michael, 1977, *Why Poor People Stay Poor: Urban Bias in World Development*, Temple Smith, London.

Longhurst, Richard, and Payne, Philip, 1979, 'Seasonal Aspects of Nutrition: Review of Evidence and Policy Implications', IDS Discussion Paper, 145, Institute of Development Studies, University of Sussex, Brighton, UK.

Longhurst, Richard (ed.), 1981a, 'Rapid Rural Appraisal: social structure and rural economy', *IDS Bulletin 12*, 4, Institute of Development Studies, University of Sussex, Brighton, UK; March.

Longhurst, Richard, 1981b, 'Research Methodology and Rural Economy in Northern Nigeria', in Longhurst, 1981a, pp. 23–31.

Lozoff, Betsy, Kamath, K. R., and Feldman, R. A., 1975, 'Infection and Disease in South Indian Families: Beliefs about Childhood Diarrhea'. *Human Organization*, 34, pp. 353–58.

McDowell, R. E., and Bove, Lea, 1977, *The Goat as a Producer of Meat*, Cornell International Agriculture Mimeo No. 56, Cornell University, Ithaca, NY 14853; November.

McDowell, R. E., and Hildebrand, P. E., 1980, *Integrated Crop and Animal Production: Making the Most of Resources Available to Small Farms in Developing Countries*, the Rockefeller Foundation, Working Papers; January.

McNamara, Robert S., 1980, *Address to the Board of Governors*, World Bank, 30 September.

Maimbo, F. J., and Fry, J., 1971, 'An Investigation into the Change in the Terms of Trade between the Rural and Urban Sectors of Zambia', *African Social Research*, 12 December.

Malik, Baljit, n.d., 'Peasant Perceptions of Poverty and Their Implications for Change Agents', photocopy, source unknown.

Mamdani, Mahmood, 1972, *The Myth of Population Control: Family, Caste and Class in an Indian Village*, Monthly Review Press, New York and London.

Marter, Alan and Honeybone, David, 1976, *The Economic Resources of Rural Households and the Distribution of Agricultural Development*, Rural Development Studies Bureau, University of Zambia, Lusaka.

Marx, Karl, 1845, 'Marx's Theses on Feuerbach (jotted down in Brussels in the spring of 1845)', in *Karl Marx Selected Works*, Vol. 1, Lawrence and Wishart Limited, London, 1942, p. 473.

Mather, R. J., and John, T. J., 1973, 'Popular Beliefs about Smallpox and

Other Infectious Diseases in South India', *Tropical and Geographical Medicine*, 25, pp. 190–96.

Mehta, Harubhai, 1979, *Legal Support Scheme for the Poor, an evaluation of one year's performance in Chota Udepur and Jetpur-Pavi Talukas*, Anand Niketan Ashram, Rangpur (Kawant), Vadodara District, Gujarat, India; May.

Mencher, Joan, 1980, 'The Lessons and Non-Lessons of Kerala: Agricultural Labourers and Poverty', *Economic and Political Weekly*, pp. 1781–1802.

Mitchell, Mark, 1982, 'Peasant Farming in the Third World', in Oxenham (ed.), *Simulations and Adult Learning for Development*, p. 28.

Moore, M. P., 1976, 'The Bureaucratic Perception of Policy Options', *IDS Bulletin 8*, 2, September, pp. 27–30.

Moore, M. P., and Wickremesinghe, G., 1980, *Agriculture and Society in the Low Country (Sri Lanka)*, Agrarian Research and Training Institute, Colombo.

Moore, Mick, 1981, 'Beyond the Tarmac Road: A Guide for Rural Poverty Watchers', in *Rapid Rural Appraisal: Social Structure and Rural Economy*, *IDS Bulletin*, 12, 4, October, pp. 47–9.

Morehouse, Ward, 1981, 'Defying Gravity: Technology and Social Justice', *Development Forum*, 9, 7, September, p. 16.

Moris, J. R., 1967, 'The Evaluation of Settlement Schemes Performance: a Sociological Appraisal', University of East Africa Social Science Conference Paper No. 430, January.

Moris, Jon R., 1977, 'The Transferability of Western Management Concepts and Programs, an East African Perspective', in Stifel *et al.* (eds), *Education and Training for Public Sector Management in Developing Countries*, pp. 73–83.

Moris, Jon R., 1981, *Managing Induced Rural Development*, International Development Institute, Indiana University, Bloomington, Indiana.

Moss, R. P., 1979, 'Soils and Crop Production in the Humid Tropics', *Area* (Institute of British Geographers), 11, 3, pp. 191–2.

Myers, Norman, n.d., 'Analysis: Why the fight against hunger is failing all across Africa', *World Environment Report*, 300 East 42nd Street, New York, NY 10017.

Myrdal, Gunnar, 1979, 'Underdevelopment and the Evolutionary Imperative', *Third World Quarterly*, Vol. 1, No. 2, April, pp. 24–42.

Myrdal, Gunnar, *Asian Drama: an Inquiry into the Poverty of Nations*, Vol. 1, Penguin Books, Harmondsworth, UK.

Netting, R., 1968, *Hill Farmers of Nigeria*, University of Washington Press, Seattle.

Norman, D. W., 1974, 'The Rationalisation of a Crop Mixture Strategy Adopted by Farmers under Indigenous Conditions: the Example of Northern Nigeria', *Journal of Development Studies*, 11, pp. 3–21.

O'Keefe, P., and Wisner, B., 1975, 'African Drought – The State of the Game', *African Environment Special Report No. 1, Problems and*

Perspectives, International African Institute, London, pp. 31–9.

O'Leary, Michael, 1980, 'Response to Drought in Kitui District, Kenya', *Disasters*, 4, 3, pp. 315–27.

Oxenham, John (ed.), 1982, 'Simulations and Adult Learning for Development', IDS Discussion Paper 172, Institute of Development Studies, University of Sussex, Brighton, UK.

OXFAM, 1980, *Field Directors' Handbook: Guidelines and information for assessing projects*, (second edition revised), OXFAM, Banbury Road, Oxford, UK.

Pacey, Arnold, 1981, *Taking Soundings for Development and Health: an approach to the information needs of rural development workers, district officials, and health services staff*, document FHE/81,2/Rev. 1, World Health Organisation, Geneva.

Palmer, Ingrid, 1976, *The New Rice in Asia: Conclusions from Four Country Studies*, United Nations Research Institute for Social Development, Geneva.

Palmer, Ingrid, 1981, 'Women's Issues and Project Appraisal', in *Rapid Rural Appraisal*, IDS Bulletin, 12, 4, October, pp. 32–9.

Panse, V. G., 1958, 'Some Comments on the Objectives and Methods of the 1960 World Census of Agriculture', *Bulletin of the International Statistical Institute*, 36, pp. 222–7. Cited in Channappa, 1977.

Pant, Niranjan, 1982, 'Close Look at SFDAs' (review of S. M. Pandey and J. S. Sodhi, *Small Farmers' Development Programme: A Study of Small Farmers' Development Agencies in Budaun, Fatehpur and Raebareilly Districts of UP*, Shri Ram Centre for Industrial Relations and Human Resources, New Delhi, 1981), *Economic and Political Weekly*, 17, 6, February 6, pp. 191–2.

Parkin, David, J., 1972, *Palms, Wine and Witnesses: Public Spirit and Private Gain in an African Farming Community*, Intertext Books. London.

Pearse, Andrew, 1980, *Seeds of Plenty, Seeds of Want: social and economic implications of the green revolution*, Clarendon Press, Oxford.

Phillipson, J., 1966, *Ecological Energetics*, Edward Arnold, London.

Pillsbury, Barbara L. K., 1979, 'Reaching the Rural Poor: Indigenous Health Practitioners Are There Already', AID Programme Evaluation Discussion Paper, No. 1, the Studies Division, Office of Evaluation, US AID, Washington; March.

Reed, W., 1970, 'The Ecology and Control of Earias'. Ph.D. Thesis, Reading University, Reading, UK.

Reynolds, Norman, and Sundar, Pushpa, 1977, 'Maharashtra's Employment Guarantee Scheme: a programme to emulate?', *Economic and Political Weekly*, 12, 29, 16 July, pp. 1149–58.

Rhoades, Robert E., 1982, *The Art of the Informal Agricultural Survey*, Social Science Department Training Document 1982–2, International Potato Center, Aptdo. 5969, Lima, Peru; March.

Richards, Paul, 1979, 'Community Environmental Knowledge in African

Rural Development', *IDS Bulletin*, 10, 2, pp. 28–36. Also in Brokensha *et al.*, 1980, pp. 183–203.

Roberts, Nigel, 1980, 'Monitoring Field Programmes (A new role for a project visitor)', typescript, June.

Rowan, John, 1976, *Ordinary Ecstasy: Humanistic Psychology in Action*, Routledge and Kegan Paul, London.

Sahlins, Marshall, 1974, *Stone Age Economics*, Tavistock Publications, London.

Sanders, John H., 1980, 'New Agricultural Technology on the Brazilian Sertao', in ICRISAT, *Proceedings of the International Workshop on Socioeconomic Constraints to Development of Semi-arid Tropical Agriculture*, 19–23 February, 1979, Hyderabad, India, pp. 73–82.

Sands, Michael, and McDowell, Robert E., 1979, *A World Bibliography on Goats*, Cornell International Agriculture Mimeo No. 56, Cornell University, Ithaca, NY 14853; November.

Santosa, Paulus, 1980, 'Community Based Nutrition Program', typescript, UPKM CD Bethesda Hospital, Jogjakarta, Indonesia.

Sarin, Sharad, and Khanna, Anand, in association with Sudas Roy, 1981, 'Management of Minor Forest Produce: an Inter-State Comprehensive Study of Tendu Leaves and Sal Seeds', typescript, Xavier Labour Relations Institute, Jamshedpur; March.

Sarin, Sharad, 1981, 'Management of Minor Forestry Produce: Perspective and Alternative Frameworks for Research and Analysis', *Indian Forester*, 107, July, pp. 397–411.

Schofield, Susan, 1974, 'Seasonal Factors Affecting Nutrition in Different Age Groups and especially Preschool Children', *Journal of Development Studies*, 11, 1; October.

Seckler, David, 1980a, 'Small but Healthy', Ford Foundation, 55 Lodi Estate, New Delhi.

Seckler, David, 1980b, 'Malnutrition: an Intellectual Odyssey', *Western Journal of Agricultural Economics*, December, pp. 219–27.

Seligman, Martin E. P., 1975, *Helplessness: on depression, development, and death*, W. H. Freeman and Company, San Francisco.

Sem, Amartya, 1981, *Poverty and Famines: an essay on entitlement and deprivation*, Clarendon Press, Oxford.

Senaratne, S. P. F., 1976, 'A Program of Micro-level Studies in Rural Sri Lanka', mimeo, no source given, 12 pages.

Senaratne, S. P. F., 1978, 'Economic Development and the Sociological Consultant: a Sri Lankan Experience', paper for the Social Science Research Council Workshop, University of Sussex, Brighton, UK; 1–2 July.

Sharpe, Kenneth Evan, 1977, *Peasant Politics: Struggle in a Dominican Village*, The Johns Hopkins University Press, Baltimore and London.

Simmons, Emmy, 1981, 'A Case Study in Food Production, Sale and Distribution', in Chambers, Longhurst and Pacey (eds), *Seasonal Dimensions to Rural Poverty*, pp. 73–80.

Sivard, Ruth Leger, 1980, *World Military and Social Expenditures 1980*, World Priorities, Box 1003, Leesburg, Virginia 22075, USA.

Snow, C. P., 1959, *The Two Cultures and the Scientific Revolution* (the Rede Lecture 1959), Cambridge University Press, Cambridge, UK.

Sokiri, Andrew R., 1972, *The Social Problems and Political Predicament of Refugees: a case study of Ibuga Refugee Settlement in West Uganda, 1967–1971*, University of Dar-es-Salaam, Student Dissertation, Political Science.

Somerset, Tony, 1982, 'Examinations Reform: the Kenya Experience', *World Bank Staff Working Paper*.

Srinivas, M. N., 1975, 'Village Studies, Participant Observation and Social Science Research in India', *Economic and Political Weekly*, 10, Nos. 33–36, pp. 1387–93.

Srinivas, M. N., Shah, A. M., and Ramaswamy, E. A. (eds), 1979, *The Fieldworker and the Field: problems and challenges in sociological investigations*, Oxford University Press, Delhi, Bombay, Calcutta, Madras.

Ssennyonga, Joseph W., 1976, 'The Cultural Dimensions of Demographic Trends', *Populi*, 3, 2, pp. 2–11.

Stavenhagen, Rodolfo, 1977, 'Basic Needs, Peasants and the Strategy for Rural Development', in Marc Nerfin (ed.), *Another Development: Approaches and Strategies*, Dag Hammarskjold Foundation, Uppsala.

Stifel, Laurence D., Coleman, James S., and Black, Joseph E. (eds), 1977, *Education and Training for Public Sector Management in Developing Countries*, The Rockefeller Foundation; March.

Stocking, Michael, and Abel, Nick, 1981, 'Ecological and Environmental Indicators for the Rapid Appraisal of Natural Resources', *Agricultural Administration*, 8, 6, November, pp. 473–84.

Sukhatme, P. V., 1977, *Malnutrition and Poverty*, Ninth Lal Bahadur Shastri Memorial Lecture, 29 January, 1977, Indian Agricultural Research Institute, New Delhi.

Swanson, Richard, 1980, 'Development Interventions and Self-Realisation Among the Gourma (Upper Volta)', in Brokensha *et al.* (eds), pp. 67–91.

Swift, Jeremy, 1979, 'Notes on Traditional Knowledge, Modern Knowledge and Rural Development', *IDS Bulletin*, 10, 2, Institute of Development Studies, University of Sussex, Brighton, UK; January.

Swift, Jeremy, 1981a, 'Labour and Subsistence in a Pastoral Economy', in Chambers, Longhurst and Pacey (eds), *Seasonal Dimensions to Rural Poverty*, pp. 80–87.

Swift, Jeremy, 1981b, 'Rapid Appraisal and Cost-effective Participatory Research in Dry Pastoral Areas of West Africa', *Agricultural Administration*, 8, 6, November, pp. 485–92.

Thomas, Gary L., 1977, *Baseline Information and Situational Overview Requisite to the Design of Integrated Rural Development Projects in Mbulu District, Tanzania*, Report Prepared for USAID/Tanzania, May.

Thomasson, G. C., 1978, 'Cultural and Psychological Variables in the Preparation and Presentation of Information to Users of Soil Resource Inventories: an Exercise in Applied Anthropology', in *Soil Resource Inventories and Development Planning*, proceedings of a workshop

held at Cornell University, 11–15 December, 1978, pp. 167–75.
Agronomy mimeo No. 79–23, Department of Agronomy, Cornell
University, Ithaca, New York 14853.

Turnbull, Colin, 1973, *The Mountain People*, Picador, Pan Books,
London.

United Nations, 1975, *Poverty, Unemployment and Development Policy:
a case study of selected issues with reference to Kerala*, Department of
Economics and Social Affairs, United Nations, New York (ST/ESA/29).

UNRISD, 1974, *The Social and Economic Implications of Large-scale
Introduction of New Varieties of Foodgrain*, United Nations Research
Institute for Social Development, Geneva.

UNRISD, 1979, *An Approach to Development Research*, United Nations
Research Institute for Social Development, Geneva.

Uphoff, Norman, and Esman, Milton, 1974, *Local Organization for Rural
Development: Analysis of Asian Experience*, Rural Local Government
Monograph 19, Rural Development Committee, Cornell University,
Ithaca NY.

Waddington, C. H., 1977, *Tools for Thought*, Paladin Frogmore, St.
Albans, UK.

Wade, Robert, and Chambers, Robert, 1980, 'Managing the Main System:
Canal Irrigation's Blind Spot', *Economic and Political Weekly*, 15, No.
39, Review of Agriculture, September, pp. A 107–A 112.

Walford, Rex, 1982, 'Simulations – the State of the Art', in Oxenham
(ed.), *Simulations and Adult Learning for Development*, pp. 10–16.

Walsh, Julia A., and Warren, Kenneth S., 1979, 'Selective Primary Health
Care: an Interim Strategy for Disease Control in Developing Countries',
The Rockefeller Foundation, 1133 Avenue of the Americas, New York,
NY 10036.

Ward, Michael, 1979, *Development Problems and Data Collection
Requirements*, Keynote Address to the Data for Development
(DFD/UNESCO/IFP) Conference, Chamrousse, May.

Warrell, David A., and Arnett, Charles, 1976, 'The Importance of Bites by
the Saw-Scaled or Carpet Viper (*Echis Carinatus*): Epidemiological
Studies in Nigeria and a Review of World Literature', *Acta Tropica* 33,
4, pp. 307–41.

Warren, Dennis M., 1980, 'Ethnoscience in Rural Development', in
Brokensha *et al.* (eds), pp. 369–81.

Weinstock, J., 1977, 'Indigenous Soil Classification', unpublished ms,
cited by Thomasson, 1978.

Weisblat, A. M., 1976, *Introduction to ADC Reprint No. 28* (reprints of
Ladejinsky, 1969a and b), Agricultural Development Council Inc.,
1290 Avenue of the Americas, New York, NY 10019; June.

Werner, Oswald, and Begishe, Kenneth Y., 1980, 'Ethnoscience in
Applied Anthopology', in Brokensha *et al.*, 1980, pp. 151–81.

Wickham, T. H., and Valera, A., 1979, 'Practices and Accountability for
Better Water Management', in Donald C. Taylor and Thomas H.
Wickham (eds), *Irrigation Policy and the Management of Irrigation*

Systems in Southeast Asia, The Agricultural Development Council, Inc., Bangkok, pp. 61–75.

World Bank, 1975, Rural Development Sector Policy Paper, World Bank, 1818 H Street NW, Washington, DC 20433.

World Bank, 1980, World Development Report 1980, World Bank, 1818 H Street NW, Washington DC 20433.

World Bank, 1981, World Development Report 1981, World Bank, 1818 H Street NW, Washington, DC 20433.

Wortman, Sterling, and Cummings Jr., Ralph W., 1978, To Feed This World: The Challenge and the Strategy, The Johns Hopkins University Press, Baltimore and London.

Index

Compiled by Helen Smith

Italicised numbers indicate tables; n after a number indicates a reference in the notes at the end of chapter.

236

Index

Index

Index

Index